T0226236

Lecture Notes in Computer Science   10836

Commenced Publication in 1973
Founding and Former Series Editors:
Gerhard Goos, Juris Hartmanis, and Jan van Leeuwen

Editorial Board

More information about this series at http://www.springer.com/series/7407

Anthony Bonato · Paweł Prałat
Andrei Raigorodskii (Eds.)

Algorithms and Models for the Web Graph

15th International Workshop, WAW 2018
Moscow, Russia, May 17–18, 2018
Proceedings

 Springer

Editors
Anthony Bonato
Department of Mathematics
Ryerson University
Toronto, ON
Canada

Paweł Prałat
Department of Mathematics
Ryerson University
Toronto, ON
Canada

Andrei Raigorodskii
Department of Discrete Mathematics
Moscow Institute of Physics and Technology
Dolgoprudny
Russia

ISSN 0302-9743   ISSN 1611-3349   (electronic)
Lecture Notes in Computer Science
ISBN 978-3-319-92870-8   ISBN 978-3-319-92871-5   (eBook)
https://doi.org/10.1007/978-3-319-92871-5

Library of Congress Control Number: 2018944417

LNCS Sublibrary: SL1 – Theoretical Computer Science and General Issues

Printed on acid-free paper

This Springer imprint is published by the registered company Springer International Publishing AG
part of Springer Nature
The registered company address is: Gewerbestrasse 11, 6330 Cham, Switzerland

Preface

The 15th Workshop on Algorithms and Models for the Web Graph (WAW 2018) took place at the Moscow Institute of Physics and Technology, Russia, May 17–18, 2018. This is an annual meeting, which is traditionally co-located with another, related, conference. WAW 2018 was co-located with the Workshop on Graphs, Networks, and Their Applications. The co-location of the two workshops provided opportunities for researchers in two different but interrelated areas to interact and to exchange research ideas. It was an effective venue for the dissemination of new results and for fostering research collaboration.

The World Wide Web has become part of our everyday life, and information retrieval and data mining on the Web are now of enormous practical interest. The algorithms supporting these activities combine the view of the Web as a text repository and as a graph, induced in various ways by links among pages, hosts and users. The aim of the workshop was to further the understanding of graphs that arise from the Web and various user activities on the Web, and stimulate the development of high-performance algorithms and applications that exploit these graphs. The workshop gathered together researchers working on graph-theoretic and algorithmic aspects of related complex networks, including social networks, citation networks, biological networks, molecular networks, and other networks arising from the Internet.

This volume contains the papers presented during the workshop. Each submission was reviewed by Program Committee members. Papers were submitted and reviewed using the EasyChair online system. The committee members accepted 11 papers.

May 2018

Anthony Bonato
Paweł Prałat
Andrei Raigorodskii

Organization

General Chairs

Andrei Z. Broder   Google Research, USA
Fan Chung Graham   University of California San Diego, USA

Organizing Committee

Anthony Bonato   Ryerson University, Canada
Paweł Prałat   Ryerson University, Canada
Andrei Raigorodskii   MIPT, Russia

Program Committee

Konstantin Avratchenkov   Inria, France
Paolo Boldi   University of Milan, Italy
Anthony Bonato   Ryerson University, Canada
Milan Bradonjic   Bell, USA
Fan Chung Graham   UC San Diego, USA
Collin Cooper   King's College London, UK
Andrzej Dudek   Western Michigan University, USA
Alan Frieze   Carnegie Mellon University, USA
Aristides Gionis   Aalto University, Finland
David Gleich   Purdue University, USA
Jeannette Janssen   Dalhousie University, Canada
Bogumil Kaminski   Warsaw School of Economics, Poland
Ravi Kumar   Google Research, USA
Silvio Lattanzi   Google Research, USA
Marc Lelarge   Inria, France
Stefano Leonardi   Sapienza University of Rome, Italy
Nelly Litvak   University of Twente, The Netherlands
Michael Mahoney   UC Berkeley, USA
Oliver Mason   NUI Maynooth, Ireland
Dieter Mitsche   Université de Nice Sophia-Antipolis, France
Peter Morters   University of Bath, UK
Tobias Mueller   Utrecht University, The Netherlands
Liudmila Ostroumova   Yandex, Russia
Pan Peng   TU Dortmund, Germany
Xavier Perez-Gimenez   University of Nebraska-Lincoln, USA
Pawel Pralat   Ryerson University, Canada
Yana Volkovich   AppNexus, USA
Stephen Young   Pacific Northwest National Laboratory, USA

Sponsoring Institutions

Microsoft Research New England, USA
Google Research, USA
Moscow Institute of Physics and Technology, Russia
Yandex, Russia
Internet Mathematics

Contents

Finding Induced Subgraphs in Scale-Free Inhomogeneous Random Graphs

Ellen Cardinaels, Johan S. H. van Leeuwaarden, and Clara Stegehuis[(⊠)]

Eindhoven University of Technology, Eindhoven, The Netherlands
`C.Stegehuis@tue.nl`

Abstract. We study the induced subgraph isomorphism problem on inhomogeneous random graphs with infinite variance power-law degrees. We provide a fast algorithm that determines for any connected graph H on k vertices if it exists as induced subgraph in a random graph with n vertices. By exploiting the scale-free graph structure, the algorithm runs in $O(nk)$ time for small values of k. We test our algorithm on several real-world data sets.

1 Introduction

The induced subgraph isomorphism problem asks whether a large graph G contains a connected graph H as an induced subgraph. When k is allowed to grow with the graph size n, this problem is NP-hard in general. For example, k-clique and k induced cycle, special cases of H, are known to be NP-hard [13,20]. For fixed k, this problem can be solved in polynomial time $O(n^k)$ by searching for H on all possible combinations of k vertices. Several randomized and non-randomized algorithms exist to improve upon this trivial way of finding H [14,25,27,29].

On real-world networks, many algorithms were observed to run much faster than predicted by the worst-case running time of algorithms. This may be ascribed to some of the properties that many real-world networks share [4], such as the power-law degree distribution found in many networks [1,8,19,28]. One way of exploiting these power-law degree distributions is to design algorithms that work well on random graphs with power-law degree distributions. For example, finding the largest clique in a network is NP-complete for general networks [20]. However, in random graph models such as the Erdős-Rényi random graph and the inhomogeneous random graph, their specific structures can be exploited to design fixed parameter tractable (FPT) algorithms that efficiently find a clique of size k [10,12] or the largest independent set [15].

In this paper, we study algorithms that are designed to perform well for the inhomogeneous random graph, a random graph model that can generate graphs with a power-law degree distribution [2,3,5,6,24,26]. The inhomogeneous random graph has a densely connected core containing many cliques, consisting of vertices with degrees $\sqrt{n \log(n)}$ and larger. In this densely connected core, the probability of an edge being present is close to one, so that it contains

© Springer International Publishing AG, part of Springer Nature 2018
A. Bonato et al. (Eds.): WAW 2018, LNCS 10836, pp. 1–15, 2018.
https://doi.org/10.1007/978-3-319-92871-5_1

many complete graphs [18]. This observation was exploited in [11] to efficiently determine whether a clique of size k occurs as a subgraph in an inhomogeneous random graph. When searching for *induced* subgraphs however, some edges are required not to be present. Therefore, searching for induced subgraphs in the entire core is not efficient. We show that a connected subgraph H can be found as an induced subgraph by scanning only vertices that are on the boundary of the core: vertices with degrees proportional to $\sqrt{n}$.

We present an algorithm that first selects the set of vertices with degrees proportional to $\sqrt{n}$, and then randomly searches for H as an induced subgraph on a subset of k of those vertices. The first algorithm we present does not depend on the specific structure of H. For general sparse graphs, the best known algorithms to solve subgraph isomorphism on 3 or 4 vertices run in $O(n^{1.41})$ or $O(n^{1.51})$ time with high probability [29]. For small values of k, our algorithm solves subgraph isomorphism on k nodes in linear time with high probability on inhomogeneous random graphs. However, the graph size needs to be very large for our algorithm to perform well. We therefore present a second algorithm that again selects the vertices with degrees proportional to $\sqrt{n}$, and then searches for induced subgraph H in a more efficient way. This algorithm has the same performance guarantee as our first algorithm, but performs much better in simulations.

We test our algorithm on large inhomogeneous random graphs, where it indeed efficiently finds induced subgraphs. We also test our algorithm on real-world network data with power-law degrees. There our algorithm does not perform well, probably due to the fact that the densely connected core of some real-world networks may not be the vertices of degrees at least proportional to $\sqrt{n}$. We then show that a slight modification of our algorithm that looks for induced subgraphs on vertices of degrees proportional to n^γ for some other value of γ performs better on real-world networks, where the value of γ depends on the specific network.

Notation. We say that a sequence of events $(\mathcal{E}_n)_{n\geq 1}$ happens with high probability (w.h.p.) if $\lim_{n\to\infty} \mathbb{P}(\mathcal{E}_n) = 1$. Furthermore, we write $f(n) = o(g(n))$ if $\lim_{n\to\infty} f(n)/g(n) = 0$, and $f(n) = O(g(n))$ if $|f(n)|/g(n)$ is uniformly bounded, where $(g(n))_{n\geq 1}$ is nonnegative. Similarly, if $\limsup_{n\to\infty} |f(n)|/g(n) > 0$, we say that $f(n) = \Omega(g(n))$ for nonnegative $(g(n))_{n\geq 1}$. We write $f(n) = \Theta(g(n))$ if $f(n) = O(g(n))$ as well as $f(n) = \Omega(g(n))$.

1.1   Model

As a random graph null model, we use the inhomogeneous random graph or hidden variable model [2,3,5,6,24,26]. Every vertex is equipped with a weight. We assume that the weights are i.i.d. samples from the power-law distribution

$$\mathbb{P}(w_i > k) = Ck^{1-\tau} \qquad (1.1)$$

for some constant C and for $\tau \in (2,3)$. Two vertices with weights w and w' are connected with probability

$$p(w, w') = \min\left(\frac{ww'}{\mu n}, 1\right), \qquad (1.2)$$

where μ denotes the mean value of the power-law distribution (1.1). Choosing the connection probability in this way ensures that the expected degree of a vertex with weight w is w.

1.2 Algorithms

We now describe two randomized algorithms that determine whether a connected graph H is an induced subgraph in an inhomogeneous random graph and find the location of such a subgraph if it exists. Algorithm 1 selects the vertices in the inhomogeneous random graph that are on the boundary of the core of the graph: vertices with degrees slightly below $\sqrt{\mu n}$. Then, the algorithm randomly divides these vertices into sets of k vertices. If one of these sets contains H as an induced subgraph, the algorithm terminates and returns the location of H. If this is not the case, then the algorithm fails. In the next section, we show that for k small enough, the probability that the algorithm fails is small. This means that H is present as an induced subgraph on vertices that are on the boundary of the core with high probability.

Algorithm 1 is similar to the algorithm in [12] designed to find cliques in random graphs. The major difference is that the algorithm to find cliques looks for cliques on all vertices with degrees larger than $\sqrt{f_1 \mu n}$ for some function f_1. This algorithm is not efficient for detecting other subgraphs than cliques, since vertices with high degrees will be connected with probability close to one.

Algorithm 1. Finding induced subgraph H (random search)

    Input : H, $G = (V, E)$, μ, $f_1 = f_1(n)$, $f_2 = f_2(n)$.
    Output: Location of H in G or fail.
1 Define $n = |V|$, $I_n = [\sqrt{f_1 \mu n}, \sqrt{f_2 \mu n}]$ and set $V' = \emptyset$.
2 **for** $i \in V$ **do**
3 | **if** $D_i \in I_n$ **then** $V' = V' \cup i$
4 **end**
5 Divide the vertices in V' randomly into $\lfloor |V'|/k \rfloor$ sets $S_1, \ldots, S_{\lfloor |V'|/k \rfloor}$.
6 **for** $j = 1, \ldots, \lfloor |V'|/k \rfloor$ **do**
7 | **if** H *is an induced subgraph on* S_j **then return** location of H
8 **end**

The following theorem gives a bound for the performance of Algorithm 1 for small values of k.

Theorem 1. *Choose* $f_1 = f_1(n) \geq 1/\log(n)$ *and* $f_1 < f_2 < 1$ *and let* $k < \log^{1/3}(n)$. *Then, with high probability, Algorithm 1 detects induced subgraph H on k vertices in an inhomogeneous random graph with n vertices and weights distributed as in (1.1) in time* $O(nk)$.

Thus, for small values of k, Algorithm 1 finds an instance of H in linear time.

A problem with parameter k is called fixed parameter tractable (FPT) if it can be solved in $f(k)n^{O(1)}$ time for some function $f(k)$, and it is called typical FPT (typFPT) if it can be solved in $f(k)n^{g(n)}$ for some function $g(n) = O(1)$ with high probability [9]. As a corollary of Theorem 1 we obtain that the induced subgraph problem on the inhomogeneous random graph is in typFPT for any subgraph H, similarly to the k-clique problem on inhomogeneous random graphs [12].

Corollary 1. *The induced subgraph problem on the inhomogeneous random graph is in typFPT.*

In theory Algorithm 1 detects any motif on k vertices in linear time for small k. However, this only holds for large values of n, which can be understood as follows. In Lemma 2, we show that $|V'| = \Theta(n^{(3-\tau)/2})$, thus tending to infinity as n grows large. However, when $n = 10^7$ and $\tau = 2.5$, this means that the size of the set V' is only proportional to $10^{1.75} = 56$ vertices. Therefore, the number of sets S_j constructed in Algorithm 1 is also small. Even though the probability of finding motif H in any such set is proportional to a constant, this constant may be small, so that for finite n the algorithm almost always fails. Thus, for Algorithm 1 to work, n needs to be large enough so that $n^{(3-\tau)/2}$ is large as well.

The algorithm can be significantly improved by changing the search for H on vertices in set V'. In Algorithm 2 we propose a search for motif H similar to the Kashtan motif sampling algorithm [21]. Rather than sampling k vertices randomly, it samples one vertex randomly, and then randomly increases the set S by adding vertices in its neighborhood. This already guarantees the vertices in list S_j to be connected, making it more likely for them to form a specific connected motif together. In particular, we expand the list S_j in such a way that the vertices in S_j are guaranteed to form a spanning tree of H as a subgraph. This is ensured by choosing the list T^H that specifies at which vertex in S_j we expand S_j by adding a new vertex. For example, if $k = 4$ and we set $T^H = [1, 2, 3]$ we first add an edge to the first vertex, then we look for a random neighbor of the previously added vertex, and then we add a random neighbor of the third added vertex. Thus, setting $T^H = [1, 2, 3]$ ensures that the set S_j contains a path of length three, whereas setting $T^H = [1, 1, 1]$ ensures that the set S_j contains a star-shaped subgraph. Depending on which subgraph H we are looking for, we can define T^H in such a way that we ensure that the set S_j at least contains a spanning tree of motif H in Step 6 of the algorithm.

The selection on the degrees ensures that the degrees are sufficiently high so that probability of finding such a connected set on k vertices is high, as well as that the degrees are sufficiently low to ensure that we do not only find complete graphs because of the densely connected core of the inhomogeneous random graph. The probability that Algorithm 2 indeed finds the desired motif H in any check is of constant order of magnitude, similar to Algorithm 1. Therefore, the performance guarantee of both algorithms is similar. However, in practice Algorithm 2 performs much better, since for finite n, k connected vertices are more likely to form a motif than k randomly chosen vertices.

Algorithm 2. Finding induced subgraph H (neighborhood search)

Input : H, $G = (V, E)$, μ, $f_1 = f_1(n)$, $f_2 = f_2(n)$, s.
Output: Location of H in G or fail.
1 Define $n = |V|$, $I_n = [\sqrt{f_1 \mu n}, \sqrt{f_2 \mu n}]$ and set $V' = \emptyset$.
2 **for** $i \in V$ **do**
3 $\quad$ **if** $D_i \in I_n$ **then** $V' = V' \cup i$
4 **end**
5 Let G' be the induced subgraph of G on vertices V'.
6 Set T^H consistently with motif H.
7 **for** $j=1,\ldots,s$ **do**
8 $\quad$ Pick a random vertex $v \in V'$ and set $S_j = v$.
9 $\quad$ **while** $|S_j| \neq k$ **do**
10 $\quad\quad$ Pick a random $v' \in N_{G'}(S_j[T^H[j]]) : v' \notin S_j$
11 $\quad\quad$ Add v' to S_j.
12 $\quad$ **end**
13 $\quad$ **if** H *is an induced subgraph on* S_j **then return** location of H
14 **end**

The following theorem shows that indeed Algorithm 2 has similar performance guarantees as Algorithm 1.

Theorem 2. *Choose* $f_1 = f_1(n) \geq 1/\log(n)$ *and* $f_1 < f_2 < 1$. *Choose* $s = \Omega(n^\alpha)$ *for some* $0 < \alpha < 1$, *such that* $s \leq n/k$. *Then, Algorithm 2 detects induced subgraph* H *on* $k < \log^{1/3}(n)$ *vertices on an inhomogeneous random graph with* n *vertices and weights distributed as in (1.1) in time* $O(nk)$ *with high probability.*

The proofs of Theorems 1 and 2 rely on the fact that for small k, any subgraph on k vertices is present in G' with high probability. This means that after the degree selection step of Algorithms 1 and 2, for small k, any motif finding algorithm can be used to find motif H on the remaining graph G', such as the Grochow-Kellis algorithm [14], the MAvisto algorithm [27] or the MODA algorithm [25]. In the proofs of Theorems 1 and 2, we show that G' has $\Theta(n^{(3-\tau)/2})$ vertices with high probability. Thus, the degree selection step reduces the problem of finding a motif H on n vertices to finding a motif on a graph with $\Theta(n^{(3-\tau)/2})$ vertices, significantly reducing the running time of the algorithms.

2   Proof of Theorems 1 and 2

We prove Theorem 1 using two lemmas. The first lemma relates the degrees of the vertices to their weights. The connection probabilities in the inhomogeneous random graph depend on the weights of the vertices. In Algorithm 1, we select vertices based on their degrees instead of their unknown weights. The following lemma shows that the weights of the vertices in V' are close to their degrees.

Lemma 1. *Degrees and weights. Fix $\varepsilon > 0$, and define $J_n = [(1-\varepsilon)\sqrt{f_1\mu n}, (1+\varepsilon)\sqrt{f_2\mu n}]$. Then, for some $K > 0$,*

$$\mathbb{P}\left(\exists i \in V' : w_i \notin J_n\right) \leq Kn\exp\left(-\frac{\varepsilon^2(1-\varepsilon)}{2(1+\varepsilon)}\sqrt{f_1\mu n}\right). \qquad (2.1)$$

Proof. Fix a vertex $i \in V$. Conditionally on the weight w_i of vertex i, $D_i \sim \text{Poi}(w_i)$ [5,16]. Then,

$$\begin{aligned}
\mathbb{P}\left(w_i < (1-\varepsilon)\sqrt{f_1\mu n},\ D_i \in I_n\right) &= \frac{\mathbb{P}\left(D_i \in I_n \mid w_i < (1-\varepsilon)\sqrt{f_1\mu n}\right)}{\mathbb{P}\left(w_i < (1-\varepsilon)\sqrt{f_1\mu n}\right)} \\
&\leq \frac{\mathbb{P}\left(D_i > \sqrt{f_1\mu n} \mid w_i = (1-\varepsilon)\sqrt{f_1\mu n}\right)}{1 - C((1-\varepsilon)\sqrt{f_1\mu n})^{1-\tau}} \\
&\leq K_1\mathbb{P}\left(D_i > \sqrt{f_1\mu n} \mid w_i = (1-\varepsilon)\sqrt{f_1\mu n}\right),
\end{aligned}$$
$$(2.2)$$

for some $K_1 > 0$. Here the first inequality follows because for Poisson random variables $\mathbb{P}\left(\text{Poi}(\lambda_1) > k\right) \leq \mathbb{P}\left(\text{Poi}(\lambda_2) > k\right)$ for $\lambda_1 < \lambda_2$. We use that by the Chernoff bound for Poisson random variables

$$\mathbb{P}\left(X > \lambda(1+\delta)\right) \leq \exp\left(-h(\delta)\delta^2\lambda/2\right), \qquad (2.3)$$

where $h(\delta) = 2((1+\delta)\ln(1+\delta) - \delta)/\delta^2$. Therefore, using that $h(\delta) \geq 1/(1+\delta)$ for $\delta \geq 0$ results in

$$\mathbb{P}\left(D_i > \sqrt{f_1\mu n} \mid w_i = (1-\varepsilon)\sqrt{f_1\mu n}\right) \leq \exp\left(-\frac{\varepsilon^2(1-\varepsilon)}{2(1+\varepsilon)}\sqrt{f_1\mu n}\right). \qquad (2.4)$$

Combining this with (2.2) and taking the union bound over all vertices then results in

$$\mathbb{P}\left(\exists i : D_i \in I_n, w_i < (1-\varepsilon)\sqrt{f_1\mu n}\right) \leq K_1n\exp\left(-\frac{\varepsilon^2(1-\varepsilon)}{2(1+\varepsilon)}\sqrt{f_1\mu n}\right). \qquad (2.5)$$

The bound for $w_i > (1+\varepsilon)\sqrt{f_2\mu n}$ follows similarly. Combining this with the fact that $f_1 < f_2$ then proves the lemma. $\qquad\square$

The second lemma shows that after deleting all vertices with degrees outside of I_n defined in Step 1 of Algorithm 1, still polynomially many vertices remain with high probability.

Lemma 2. *Polynomially many nodes remain. There exists $\gamma > 0$ such that*

$$\mathbb{P}\left(|V'| < \gamma n^{(3-\tau)/2}\right) \leq 2\exp\left(-\Theta(n^{(3-\tau)/2})\right). \qquad (2.6)$$

Proof. Let $\mathcal{E}$ denote the event that all vertices $i \in V'$ satisfy $w_i \in J_n$ for some $\varepsilon > 0$, with J_n as in Lemma 1. Let W' be the set of vertices with weights in J_n. Under the event $\mathcal{E}$, $|V'| \leq |W'|$. Then, by Lemma 1

$$\mathbb{P}\left(|V'| < \gamma n^{(3-\tau)/2}\right) \leq \mathbb{P}\left(|W'| < \gamma n^{(3-\tau)/2}\right) + Kn\exp\left(-\frac{\varepsilon^2(1-\varepsilon)}{2(1+\varepsilon)}\sqrt{f_1\mu n}\right). \qquad (2.7)$$

Furthermore,

$$\mathbb{P}\left(w_i \in J_n\right) = C((1-\varepsilon)\sqrt{f_1\mu n})^{1-\tau} - C((1+\varepsilon)\sqrt{f_2\mu n})^{1-\tau} \geq c_1(\sqrt{\mu n})^{1-\tau} \tag{2.8}$$

for some constant $c_1 > 0$ because $f_1 < f_2$. Thus, each of the n vertices is in set W' independently with probability at least $c_1(\sqrt{\mu n})^{1-\tau}$. Choose $0 < \gamma < c_1$. Applying the multiplicative Chernoff bound then shows that

$$\mathbb{P}\left(|W'| < \gamma n^{(3-\tau)/2}\right) \leq \exp\left(-\frac{(c_1-\gamma)^2}{2c_1} n^{(3-\tau)/2}\right), \tag{2.9}$$

which proves the lemma together with (2.7) and the fact that $\sqrt{f_1\mu n} = \Omega(n^{(3-\tau)/2})$ for $\tau \in (2,3)$. $\qquad\square$

We now use these lemmas to prove Theorem 1.

Proof of Theorem 1. We condition on the event that V' is of polynomial size (Lemma 2) and that the weights are within the constructed lower and upper bounds (Lemma 1), since both events occur with high probability. This bounds the edge probability between any pair of nodes i and j in V' as

$$p_{ij} < \min\left(\frac{(1+\varepsilon)\sqrt{f_2\mu n}(1+\varepsilon)\sqrt{f_2\mu n}}{\mu n}, 1\right) = f_2(1+\varepsilon)^2, \tag{2.10}$$

so that $p_{ij} \leq p_+ = c_1 < 1$ if we choose ε small enough. Similarly,

$$p_{ij} > \min\left(\frac{(1-\varepsilon)^2\sqrt{f_1\mu n}^2}{\mu n}\right) = \Theta\left(\frac{1}{\log(n)}\right), \tag{2.11}$$

by our choice of f_1, so that $p_{ij} \geq p_- = c_2/\log(n)$. Let $E := |E_H|$ be the number of edges in H. We upper bound the probability of not finding H in one of the partitions of size k of V' as $1 - p_-^E(1-p_+)^{\binom{k}{2}-E}$. Since all partitions are disjoint we can upper bound the probability of not finding H in any of the partitions as

$$\mathbb{P}\left(H \text{ not in the partitions}\right) \leq \left(1 - p_-^E(1-p_+)^{\binom{k}{2}-E}\right)^{\lceil \frac{|V'|}{k} \rceil}. \tag{2.12}$$

Using that $E \leq k^2$, $\binom{k}{2} - E \leq k^2$ and that $1 - x \leq e^{-x}$ results in

$$\mathbb{P}\left(H \text{ not in the partitions}\right) \leq \exp\left(-p_-^{k^2}(1-p_+)^{k^2}\left\lceil\frac{|V'|}{k}\right\rceil\right). \tag{2.13}$$

Since $|V'| = \Theta\left(n^{\frac{3-\tau}{2}}\right)$, $\lceil|V'|/k\rceil \geq dn^{\frac{3-\tau}{2}}/k$ for some constant $d > 0$. We fill in the expressions for p_- and p_+, with $c_3 > 0$ a constant

$$\mathbb{P}\left(H \text{ not in the partitions}\right) \leq \exp\left(-\frac{dn^{\frac{3-\tau}{2}}}{k}\left(\frac{c_3}{\log n}\right)^{k^2}\right). \tag{2.14}$$

Now apply that $k \leq \log^{\frac{1}{3}}(n)$. Then

$$
\mathbb{P}\left(H \text{ not in the partitions}\right) \leq \exp\left(-\frac{dn^{\frac{3-\tau}{2}}}{\log^{\frac{1}{3}}n}\left(\frac{c_3}{\log n}\right)^{\log^{\frac{2}{3}}n}\right) \tag{2.15}
$$
$$
\leq \exp\left(-dn^{\frac{3-\tau}{2}-o(1)}\right).
$$

Hence, the inner expression grows polynomially such that the probability of not finding H in one of the partitions is negligibly small. The running time of the partial search is given by

$$
\frac{|V'|}{k}\binom{k}{2} \leq \frac{n}{k}\binom{k}{2} \leq nk \leq ne^{k^4}, \tag{2.16}
$$

which concludes the proof for $k \leq \log^{1/3}(n)$.   □

Proof of Corollary 1. If $k > \log^{\frac{1}{3}}(n)$, we can determine whether H is an induced subgraph by exhaustive search in time

$$
\binom{n}{k}\binom{k}{2} \leq \frac{n^k}{k}\frac{k(k-1)}{2} \leq kn^k \leq ke^{k^4} \leq ne^{k^4}, \tag{2.17}
$$

since for all sets of k vertices the presence or absence of $\binom{k}{2}$ edges needs to be checked. For $k \leq \log^{\frac{1}{3}}(n)$, Theorem 1 shows that the induced subgraph isomorphism problem can be solved in time $nk \leq ne^{k^4}$. Thus, with high probability the induced subgraph isomorphism problem can be solved in ne^{k^4} time, which proves that it is in typFPT.   □

Proof of Theorem 2. The proof of Theorem 2 is very similar to the proof of Theorem 1. The only way Algorithm 2 differs from Algorithm 1 is in the selection of the sets S_j. As in the previous theorem, we condition on the event that $|V'| = \Theta(n^{(3-\tau)/2})$ (Lemma 2) and that the weights of the vertices in G' are bounded as in Lemma 1.

The graph G' constructed in Step 5 of Algorithm 2 then consists of $\Theta(n^{(3-\tau)/2})$ vertices. Furthermore, by the bound (2.11) on the connection probabilities of all vertices in G', the expected degree of a vertex i in G' satisfies $\mathbb{E}[D_{i,G'}] = \Omega(n^{(3-\tau)/2}/\log(n))$. We can use similar arguments as in Lemma 1 to show that $D_{i,G'} = \Omega(n^{(3-\tau)/2}/\log(n))$ with high probability for all vertices in G'. Since G' consists of $\Theta(n^{(3-\tau)/2})$ vertices, $D_{i,G'} = O(n^{(3-\tau)/2})$ as well. This means that for $k < \log^{\frac{1}{3}}(n)$, Steps 8–11 are able to find a connected subgraph on k vertices with high probability.

We now compute the probability that S_j is disjoint with the previous $j-1$ constructed sets. The probability that the first vertex does not overlap with the previous sets is given by $1 - jk/|V'|$, since that vertex is chosen uniformly at random. The second vertex is chosen in a size-biased manner, since it is chosen by following a random edge. The probability that vertex i is added can therefore be bounded as

$$\mathbb{P}\left(\text{vertex } i \text{ is added}\right) = \frac{D_{i,G'}}{\sum_{s=1}^{|V'|} D_{s,G'}} \leq \frac{M \log(n)}{|V'|} \qquad (2.18)$$

for some constant $M > 0$ by the conditions on the degrees. Therefore, the probability that S_j does not overlap with one of the previously chose jk vertices can be bounded from below by

$$\mathbb{P}\left(S_j \text{ does not overlap with previous sets}\right) \geq \left(1 - \frac{kj}{|V'|}\right)\left(1 - \frac{Mkj \log(n)}{|V'|}\right)^{k-1}. \qquad (2.19)$$

Thus, the probability that all j sets do not overlap can be bounded as

$$\mathbb{P}\left(S_j \cap S_{j-1} \cdots \cap S_1 = \emptyset\right) \geq \left(1 - \frac{Mkj \log(n)}{|V'|}\right)^{jk}, \qquad (2.20)$$

which tends to one when $jk = o(n^{(3-\tau)/4})$. Let s_{dis} denote the number of disjoint sets out of the s sets constructed in Algorithm 2. Then, when $s = \Omega(n^\alpha)$ for some $\alpha > 0$, $s_{\text{dis}} > n^\beta$ for some $\beta > 0$ with high probability, because $k < \log^{1/3}(n)$.

The probability that H is present as an induced subgraph is bounded similarly as in Theorem 1. We already know that $k - 1$ edges are present. For all other $E - (k - 1)$ edges of H, and all $\binom{k}{2} - E$ edges that are not present in H, we can again use (2.10) and (2.11) to bound on the probability of edges being present or not being present between vertices in V'. Therefore, we can bound the probability that H is not found similarly to (2.13) as

$$\mathbb{P}\left(H \text{ not in the partitions}\right) \leq \mathbb{P}\left(H \text{ not in the disjoint partitions}\right)$$

$$\leq \exp\left(-p_-^{k^2}(1 - p_+)^{k^2} s_{\text{dis}}\right).$$

Because $s_{\text{dis}} > n^\beta$ for some $\beta > 0$, this term tends to zero exponentially. The running time of the partial search can be bounded similarly to (2.16) as

$$s\binom{k}{2} \leq sk^2 = O(nk), \qquad (2.21)$$

where we used that $s \leq n/k$. $\qquad \square$

3   Experimental Results

Fig. 1 shows the fraction of times Algorithm 1 succeeds to find a cycle of size k in an inhomogeneous random graph on 10^7 vertices. Even though for large n Algorithm 1 should find an instance of a cycle of size k in step 7 of the algorithm with high probability, we see that Algorithm 1 never succeeds in finding one. This is because of the finite size effects discussed before.

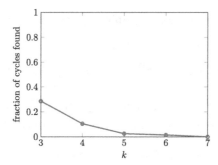

Fig. 1. The fraction of times step 7 in Algorithm 1 succeeds to find a cycle of length k on an inhomogeneous random graph with $n = 10^7$, averaged over 500 network samples with $f_1 = 1/\log(n)$ and $f_2 = 0.9$.

Figure 2a also plots the fraction of times Algorithm 2 succeeds to find a cycle. We set the parameter $s = 10000$ so that the algorithm fails if the algorithm does not succeed to detect motif H after executing step 13 of Algorithm 2 10000 times. Because s gives the number of attempts to find H, increasing s may increase the success probability of Algorithm 2 at the cost of a higher running time. However, in Fig. 2b we see that for small values of k, the mean number of times Step 13 is executed when the algorithm succeeds is much lower than 10000, so that increasing s in this experiment probably only has a small effect on the success probability. We see that Algorithm 2 outperforms Algorithm 1. Figure 2b also shows that the number of attempts needed to detect a cycle of length k is small for $k \leq 6$. For larger values of k the number of attempts increases. This can again be ascribed to the finite size effects that cause the set V' to be small, so that large motifs may not be present on vertices in set V'. We also plot the success probability when using different values of the functions f_1 and f_2. When only the lower bound f_1 on the vertex degrees is used, as in [11], the success probability of the algorithm decreases. This is because the set V' now contains many high degree vertices that are much more likely to form clique motifs than cycles or other connected motifs on k vertices. This makes $f_2 = \infty$ a very efficient bound for detecting clique motifs [11]. For the cycle motif however, we see in Fig. 2b that more checks are needed before a cycle is detected, and in some cases the cycle is not detected at all.

Setting $f_1 = 0$ and $f_2 = \infty$ is also less efficient, as Fig. 2a shows. In this situation, the number of attempts needed to find a cycle of length k is larger than for Algorithm 2 for $k \leq 6$.

3.1   Real Network Data

We now check Algorithm 2 on four real-world networks with power-law degrees: a Wikipedia communication network [22], the Gowalla social network [22], the Baidu online encyclopedia [23] and the Internet on the autonomous systems level [22]. Table 1 presents several statistics of these scale-free data sets. Fig. 3

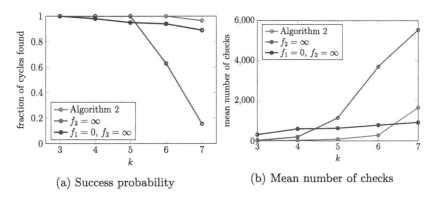

(a) Success probability   (b) Mean number of checks

Fig. 2. Results of Algorithm 2 on an inhomogeneous random graph with $n = 10^7$ for detecting cycles of length k. The parameters are chosen as $s = 10000$, $f_1 = 1/\log(n)$, $f_2 = 0.9$. The values are averaged over 500 generated networks.

shows the fraction of runs where Algorithm 2 finds a cycle as an induced subgraph. We see that for the Wikipedia social network in Fig. 3a, Algorithm 2 is more efficient than looking for cycles among all vertices in the network. For the Baidu online encyclopedia in Fig. 3c however, we see that Algorithm 2 performs much worse than looking for cycles among all possible vertices. In the other two network data sets in Figs. 3b and d the performance on the reduced vertex set and the original vertex set is almost the same. Figure 4 shows that in general, Algorithm 2 indeed seems to finish in fewer steps than when using the full vertex set. However, as Fig. 4c shows, for larger values of k the algorithm fails almost always.

Table 1. Statistics of the data sets: the number of vertices n, the number of edges E, and the power-law exponent τ fitted by the method of [7].

	n	E	τ
Wikipedia	2,394,385	5,021,410	2.46
Gowalla	196,591	950,327	2.65
Baidu	2,141,300	17,794,839	2.29
AS-Skitter	1,696,415	11,095,298	2.35

These results show that while Algorithm 2 is efficient on inhomogeneous random graphs, it may not always be efficient on real-world data sets. This is not surprising, because there is no reason why the vertices of degrees proportional to $\sqrt{n}$ should behave like an Erdős-Rényi random graph, like in the inhomogeneous random graph. We therefore investigate whether selecting vertices with degrees in $I_n = [(\mu n)^\gamma / \log(n), (\mu n)^\gamma]$ for some other value of γ in Algorithm 2 leads to a better performance. Figures 3 and 4 show for every data set one particular

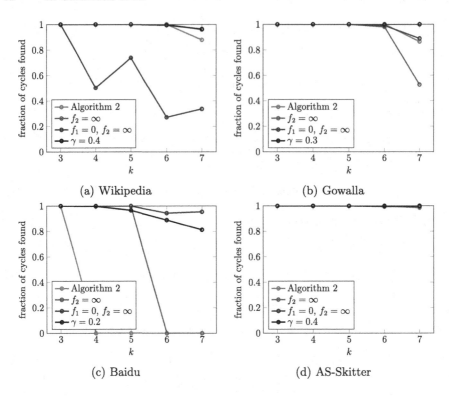

Fig. 3. The fraction of times Algorithm 2 succeeds to find a cycle on four large network data sets for detecting cycles of length k. The parameters are chosen as $s = 10000$, $f_1 = 1/\log(n)$, $f_2 = 0.9$. The black line uses Algorithm 2 on vertices of degrees in $I_n = [(\mu n)^\gamma / \log(n), (\mu n)^\gamma]$. The values are averaged over 500 runs of Algorithm 2.

value of γ that works well. For the Gowalla, Wikipedia and Autonomous systems network, this leads to a faster algorithm to detect cycles. Only for the Baidu network other values of γ do not improve upon randomly selecting from all vertices. This indicates that for most networks, cycles do appear mostly on degrees with specific orders of magnitude, making it possible to sample these cycles faster. Unfortunately, these orders of magnitude may be different for different networks. Across all four networks, the best value of γ seems to be smaller than the value of 0.5 that is optimal for the inhomogeneous random graph.

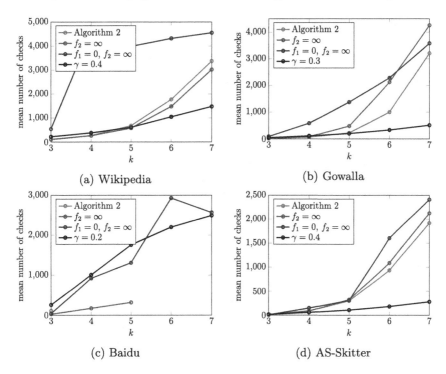

Fig. 4. The number of times step 12 of Algorithm 2 is invoked when the algorithm does not fail on four large network data sets for detecting cycles of length k. The parameters are chosen as $s = 10000$, $f_1 = 1/\log(n)$, $f_2 = 0.9$. The black line uses Algorithm 2 on vertices of degrees in $I_n = [(\mu n)^\gamma/\log(n), (\mu n)^\gamma]$. The values are averaged over 500 runs of Algorithm 2.

4   Conclusion

We presented an algorithm which solves the induced subgraph problem on inhomogeneous random graphs with infinite variance power-law degrees in time $O(ne^{k^4})$ with high probability as n grows large. This algorithm is based on the observation that for fixed k, any subgraph is present on k vertices with degrees slightly smaller than $\sqrt{\mu n}$ with positive probability. Therefore, the algorithm first selects vertices with those degrees, and then uses a random search method to look for the induced subgraph on those vertices.

We show that this algorithm performs well on simulations of inhomogeneous random graphs. Its performance on real-world data sets varies for different data sets. This indicates that the degrees that contain the most induced subgraphs of size k in real-world networks may not be close to $\sqrt{n}$. We then show that on these data sets, it may be more efficient to find induced subgraphs on degrees proportional to n^γ for some other value of γ. The value of γ may be different for different networks.

Our algorithm exploits that induced subgraphs are likely formed among $\sqrt{\mu n}$-degree vertices. However, certain subgraphs may occur more frequently on vertices of other degrees [17]. For example, star-shaped subgraphs on k vertices appear more often on one vertex with degree much higher than $\sqrt{\mu n}$ corresponding to the middle vertex of the star, and $k-1$ lower-degree vertices corresponding to the leafs of the star [17]. An interesting open question is whether there exist better degree-selection steps for specific subgraphs than the one used in Algorithms 1 and 2.

Acknowledgements. The work of JvL and CS was supported by NWO TOP grant 613.001.451. The work of JvL was further supported by the NWO Gravitation Networks grant 024.002.003, an NWO TOP-GO grant and by an ERC Starting Grant.

References

1. Albert, R., Jeong, H., Barabási, A.L.: Internet: diameter of the world-wide web. Nature **401**(6749), 130–131 (1999)
2. Boguñá, M., Pastor-Satorras, R.: Class of correlated random networks with hidden variables. Phys. Rev. E **68**, 036112 (2003)
3. Bollobás, B., Janson, S., Riordan, O.: The phase transition in inhomogeneous random graphs. Random Struct. Algorithms **31**(1), 3–122 (2007)
4. Brach, P., Cygan, M., Łacki, J., Sankowski, P.: Algorithmic complexity of power law networks. In: Proceedings of the Twenty-Seventh Annual ACM-SIAM Symposium on Discrete Algorithms, SODA 2016, pp. 1306–1325. Society for Industrial and Applied Mathematics, Philadelphia (2016)
5. Britton, T., Deijfen, M., Martin-Löf, A.: Generating simple random graphs with prescribed degree distribution. J. Stat. Phys. **124**(6), 1377–1397 (2006)
6. Chung, F., Lu, L.: The average distances in random graphs with given expected degrees. Proc. Natl. Acad. Sci. USA **99**(25), 15879–15882 (2002) (electronic)
7. Clauset, A., Shalizi, C.R., Newman, M.E.J.: Power-law distributions in empirical data. SIAM Rev. **51**(4), 661–703 (2009)
8. Faloutsos, M., Faloutsos, P., Faloutsos, C.: On power-law relationships of the internet topology. ACM SIGCOMM Comput. Commun. Rev. **29**, 251–262 (1999)
9. Fountoulakis, N., Friedrich, T., Hermelin, D.: On the average-case complexity of parameterized clique. arXiv:1410.6400v1 (2014)
10. Fountoulakis, N., Friedrich, T., Hermelin, D.: On the average-case complexity of parameterized clique. Theor. Comput. Sci. **576**, 18–29 (2015)
11. Friedrich, T., Krohmer, A.: Cliques in hyperbolic random graphs. In: INFOCOM Proceedings 2015, pp. 1544–1552. IEEE (2015)
12. Friedrich, T., Krohmer, A.: Parameterized clique on inhomogeneous random graphs. Disc. Appl. Math. **184**, 130–138 (2015)
13. Garey, M.R., Johnson, D.S., Garey, M.R.: Computers and Intractability: A Guide to the Theory of NP-Completeness. W H FREEMAN & CO (2011)
14. Grochow, J.A., Kellis, M.: Network motif discovery using subgraph enumeration and symmetry-breaking. In: RECOMB, pp. 92–106 (2007)
15. Heydari, H., Taheri, S.M.: Distributed maximal independent set on inhomogeneous random graphs. In: 2017 2nd Conference on Swarm Intelligence and Evolutionary Computation (CSIEC). IEEE, March 2017

16. van der Hofstad, R.: Random Graphs and Complex Networks, vol. 1. Cambridge University Press, Cambridge (2017)
17. van der Hofstad, R., van Leeuwaarden, J.S.H., Stegehuis, C.: Optimal subgraph structures in scale-free networks. arXiv:1709.03466 (2017)
18. Janson, S., Łuczak, T., Norros, I.: Large cliques in a power-law random graph. J. Appl. Probab. **47**(04), 1124–1135 (2010)
19. Jeong, H., Tombor, B., Albert, R., Oltvai, Z.N., Barabási, A.L.: The large-scale organization of metabolic networks. Nature **407**(6804), 651–654 (2000)
20. Karp, R.M.: Reducibility among combinatorial problems. In: Miller, R.E., Thatcher, J.W., Bohlinger, J.D. (eds.) Complexity of Computer Computations. The IBM Research Symposia Series, pp. 85–103. Springer, Boston (1972). https://doi.org/10.1007/978-1-4684-2001-2_9
21. Kashtan, N., Itzkovitz, S., Milo, R., Alon, U.: Efficient sampling algorithm for estimating subgraph concentrations and detecting network motifs. Bioinformatics **20**(11), 1746–1758 (2004)
22. Leskovec, J., Krevl, A.: SNAP Datasets: Stanford large network dataset collection (2014). http://snap.stanford.edu/data. Accessed 14 Mar 2017
23. Niu, X., Sun, X., Wang, H., Rong, S., Qi, G., Yu, Y.: Zhishi.me - weaving chinese linking open data. In: Aroyo, L., Welty, C., Alani, H., Taylor, J., Bernstein, A., Kagal, L., Noy, N., Blomqvist, E. (eds.) ISWC 2011. LNCS, vol. 7032, pp. 205–220. Springer, Heidelberg (2011). https://doi.org/10.1007/978-3-642-25093-4_14
24. Norros, I., Reittu, H.: On a conditionally poissonian graph process. Adv. Appl. Probab. **38**(01), 59–75 (2006)
25. Omidi, S., Schreiber, F., Masoudi-Nejad, A.: MODA: an efficient algorithm for network motif discovery in biological networks. Genes Genetic Syst. **84**(5), 385–395 (2009)
26. Park, J., Newman, M.E.J.: Statistical mechanics of networks. Phys. Rev. E **70**, 066117 (2004)
27. Schreiber, F., Schwobbermeyer, H.: MAVisto: a tool for the exploration of network motifs. Bioinformatics **21**(17), 3572–3574 (2005)
28. Vázquez, A., Pastor-Satorras, R., Vespignani, A.: Large-scale topological and dynamical properties of the internet. Phys. Rev. E **65**, 066130 (2002)
29. Williams, V.V., Wang, J.R., Williams, R., Yu, H.: Finding four-node subgraphs in triangle time. In: Proceedings of the Twenty-Sixth Annual ACM-SIAM Symposium on Discrete Algorithms, SODA 2015, pp. 1671–1680. Society for Industrial and Applied Mathematics, Philadelphia (2015)

The Asymptotic Normality of the Global Clustering Coefficient in Sparse Random Intersection Graphs

Mindaugas Bloznelis[1]([⊠]) and Jerzy Jaworski[2]

[1] Institute of Computer Science, Vilnius University, 03225 Vilnius, Lithuania
`mindaugas.bloznelis@mif.vu.lt`
[2] Faculty of Mathematics and Computer Science, Adam Mickiewicz University,
61-614 Poznań, Poland
`jaworski@amu.edu.pl`

Abstract. We establish the asymptotic normality of the global clustering coefficient in sparse uniform random intersection graphs.

Keywords: Clustering coefficient · Asymptotic normality
Random intersection graph

1 Introduction

The global clustering coefficient of a finite graph $\mathcal{G}$ is the ratio $C_{\mathcal{G}} = 3N_\Delta/N_\vee$, where N_Δ is the number of triangles and $N_\vee$ is the number of paths of length 2. Equivalently, $C_{\mathcal{G}}$ represents the probability that a randomly selected path of length 2 induces triangle in $\mathcal{G}$. The global clustering coefficient is a commonly used network characteristic, assessing the strength of the statistical association between neighboring adjacency relations. For example, in a social network the tendency of linking actors which have a common neighbor is reflected by a non-negligible value of the global clustering coefficient.

Clustering in a social network can be explained by an auxiliary bipartite structure: each actor is prescribed a collection of attributes and any two actors sharing a common attribute have high chances of being adjacent, cf. [8]. The respective random intersection graph (RIG) on the vertex set $V = \{v_1, \ldots, v_n\}$ and with the auxiliary attribute set $W = \{w_1, \ldots, w_m\}$ defines adjacency relations with the help of a random bipartite graph H linking actors (=vertices) to attributes: two actors are adjacent in RIG if they have a common neighbour in H. We mention that RIG admits non-vanishing tunable global clustering coefficient, power-law degrees and short typical distances, see e.g., [4].

In this note we consider the uniform random intersection graph $G(n, m, r)$, where every vertex $v_i \in V$ is prescribed a random subset $S_i = S(v_i) \subset W$ of size r and two vertices v_i, v_j are declared adjacent (denoted $v_i \sim v_j$) whenever $S_i \cap S_j \neq \emptyset$. We assume that the sets $S_1, \ldots, S_n$ are independent. (The respective random bipartite graph H is drawn uniformly at random from the class of bipartite

© Springer International Publishing AG, part of Springer Nature 2018
A. Bonato et al. (Eds.): WAW 2018, LNCS 10836, pp. 16–29, 2018.
https://doi.org/10.1007/978-3-319-92871-5_2

graphs with the property that each actor $v_i \in V$ has exactly r neighbours in W.) The uniform random intersection graph has been widely studied in the literature mainly as a model of secure wireless sensor network that uses random predistribution of keys, see [5,14]. We denote for short $G = G(n, m, r)$ and by $\mathcal{G}$ we denote the instance (realization) of the random graph G.

We consider large random intersection graphs, where $r^2 = o(m)$ as $m, n \to +\infty$. In this case the edge probability is, see (53),

$$p_e = \mathbf{P}(v_i \sim v_j) = r^2 m^{-1} + O(r^4 m^{-2}). \tag{1}$$

For us the most interesting range of parameters n, m, r is defined by the approximate relation

$$m \approx cnr^2, \tag{2}$$

where $c > 0$ is an arbitrary constant. In this case we obtain a sparse random graph, where the expected number of edges $\binom{n}{2}p_e \approx n/(2c)$ scales as n.

Before formulating our results we introduce some notation. Given a vertex triple v_i, v_j, v_k, let $\Delta_{i,j,k}$ and p_Δ denote the indicator and the probability of the event that the vertex triple induces a triangle in G. Similarly, $\vee_{ijk}$ and $p_\vee$ denote the indicator and probability that G contains the path $v_i \sim v_j \sim v_k$ (we call such a path a cherry). The total number of triangles N_Δ and cherries $N_\vee$ are

$$N_\Delta = N_\Delta(S_1, \ldots, S_n) = \sum_{\{i,j,k\} \subset [n]} \Delta_{i,j,k}, \tag{3}$$

$$N_\vee = N_\vee(S_1, \ldots, S_n) = \sum_{\{i,j,k\} \subset [n]} \left(\vee_{ijk} + \vee_{jki} + \vee_{kij} \right).$$

Denote

$$\bar{N}_\Delta = N_\Delta - \mathbf{E}N_\Delta, \bar{N}_\vee = N_\vee - \mathbf{E}N_\vee, \sigma_\Delta^2 = \mathbf{E}\bar{N}_\Delta^2, \sigma_\vee^2 = \mathbf{E}\bar{N}_\vee^2, \sigma_{\Delta\vee} = \mathbf{E}(\bar{N}_\Delta \bar{N}_\vee).$$

We start our analysis with an evaluation of the first and second moments of the subgraph counts N_Δ and $N_\vee$.

Lemma 1. *Let $m, n \to +\infty$. Assume that $r \geq 2$ and $r^3 = O(m)$. We have*

$$\mathbf{E}N_\Delta = \binom{n}{3}p_\Delta, \quad p_\Delta = \frac{r^3}{m^2} + \frac{r^6}{m^3} + O\left(\frac{r^5}{m^3}\right), \tag{4}$$

$$\mathbf{E}N_\vee = 3\binom{n}{3}p_\vee, \quad p_\vee = p_e^2 = \frac{r^4}{m^2} - \frac{r^4(r-1)^2}{m^3} + \frac{r^4(r-1)^4}{4m^4} + O\left(\frac{r^8}{m^4}\right), \tag{5}$$

$$\sigma_\Delta^2 = (n-2)^2 \binom{n}{2} \mathbf{E}g_{\Delta 1,2}^2 + \binom{n}{3} \mathbf{E}h_{\Delta 1,2,3}^2, \tag{6}$$

$$\sigma_\vee^2 = (n-2)^2 \binom{n}{2} \mathbf{E}g_{\vee 1,2}^2 + \binom{n}{3} \mathbf{E}h_{\vee 1,2,3}^2, \tag{7}$$

$$\sigma_{\Delta\vee} = (n-2)^2 \binom{n}{2} \mathbf{E}(g_{\Delta 1,2} g_{\vee 1,2}) + \binom{n}{3} \mathbf{E}(h_{\Delta 1,2,3} h_{\vee 1,2,3}). \tag{8}$$

The random variables $g_{\Delta 1,2}$, $h_{\Delta 1,2,3}$ and $g_{\vee 1,2}$, $h_{\vee 1,2,3}$ define the Hoeffding decomposition of $\bar{N}_\Delta$ and $\bar{N}_\vee$, see (12). Their second moments entering (6), (7), (8) are evaluated in (25), (26) and (31), (32) and (39), (40) respectively.

We note that (4) and (5) imply that the "theoretical clustering coefficient"

$$\mathbf{P}\left(\Delta_{i,j,k}\big|\vee_{ijk}\right) = \frac{p_\Delta}{p_\vee} = \frac{\mathbf{E}(3N_\Delta)}{\mathbf{E}N_\vee} \approx \frac{1}{r} \qquad \text{as} \qquad n, m \to +\infty.$$

Therefore, in order to have a non-vanishing global clustering coefficient we need r to be bounded as $n, m \to +\infty$, cf. [3,13]. But we may still expect the asymptotic normality of $\sigma_\Delta^{-1}\bar{N}_\Delta$ and $\sigma_\vee^{-1}\bar{N}_\vee$ even for $r \to \infty$ as $n, m \to +\infty$. Indeed, assuming (2) we obtain from (4) for $r^3 = o(m)$ that

$$\mathbf{E}N_\Delta \approx \frac{m}{6c^3r^3}\left(1 + \frac{r^3}{m}\right) \to +\infty \qquad \text{as} \qquad n, m \to +\infty. \qquad (9)$$

Hence, for $r^3 = o(m)$ we can expect the asymptotic normality of $\sigma_\Delta^{-1}\bar{N}_\Delta$. For larger r such that $m = O(r^3)$ and $r^2 = o(m)$, the identity $\mathbf{E}N_\Delta = \binom{n}{3}p_\Delta$ combined with (2) and the bound $p_\Delta = O(r^3m^{-2}+r^6m^{-3})$ implies $\mathbf{E}N_\Delta = O(1)$. The latter bound rules out the asymptotic normality of $\sigma_\Delta^{-1}\bar{N}_\Delta$. We refer to Lemma 4 and the remark following it for various bounds on p_Δ.

Our main result, Theorem 2 below gives sufficient conditions for the asymptotic normality of $C_\mathcal{G}$ as $n, m \to +\infty$. We derive the asymptotic normality of $C_\mathcal{G}$ from a related asymptotic normality result for the bivariate vector of subgraph counts $(N_\Delta, N_\vee)$.

Theorem 1. *Let $\alpha, \beta > 0$. Let $m, n \to +\infty$. Assume that $\alpha \le m/n \le \beta$. Assume that $r \ge 2$ and $r = O(1)$. Suppose that the ratio $\sigma_{\Delta\vee}/(\sigma_\Delta\sigma_\vee)$ converges to a limit. We denote the limit $\varkappa$. The random vector $\left(\sigma_\Delta^{-1}\bar{N}_\Delta, \sigma_\vee^{-1}\bar{N}_\vee\right)$ converges in distribution to a Gaussian random vector (η_1, η_2), where $\mathbf{E}\eta_j = 0$, $\mathbf{E}\eta_j^2 = 1$, $j = 1, 2$, and $\mathbf{E}\eta_1\eta_2 = \varkappa$.*

An immediate consequence of Theorem 1 is the asymptotic normality of the global clustering coefficient $C_\mathcal{G}$.

Theorem 2. *Let $r \ge 2$ and $\beta > 0$. Let $m, n \to +\infty$. Assume that $m/n \to \beta$. Then the ratio $\sigma_{\Delta\vee}/(\sigma_\Delta\sigma_\vee)$ converges to a limit. We denote the limit $\varkappa$. The random variable*

$$\sigma^{-1}\left(C_\mathcal{G} - 3\frac{\mathbf{E}N_\Delta}{\mathbf{E}N_\vee}\right)$$

converges in distribution to the standard normal random variable. Here

$$\sigma^2 = 9\left(\frac{\mathbf{E}N_\Delta}{\mathbf{E}N_\vee}\right)^2\left(\left(\frac{\sigma_\Delta}{\mathbf{E}N_\Delta}\right)^2 + \left(\frac{\sigma_\vee}{\mathbf{E}N_\vee}\right)^2 - 2\varkappa\frac{\sigma_\Delta\sigma_\vee}{\mathbf{E}N_\Delta\mathbf{E}N_\vee}\right).$$

We remark that the asymptotic normality of subgraph counts like N_Δ, $N_\vee$ and their derivatives such as $C_\mathcal{G}$ provide a useful tool for statistical inference in network analysis, see e.g., [12]. Results of Theorems 1 and 2 seem to be new. We are not aware of an earlier work on the asymptotic normality of the global clustering coefficient in *sparse* random graphs. A related problem of Poissonian approximation of the number of cliques in random intersection graphs has been addressed in [9].

Future Work. We envisage the extension of the techniques developed in the present paper to more general sparse random intersection graphs and to the counts of subgraphs of arbitrary, but finite size.

2 Proofs

In the proof we combine Hoeffding's decomposition and Stein's method. In a bit different context a similar approach has been used in [2], see also [7].

The section is organized as follows. We first collect necessary notation. Then we construct Hoeffding decompositions of $\bar{N}_\Delta$, $\bar{N}_\vee$ and evaluate variances of various parts of the decompositions. Next we briefly outline our approach to the asymptotic normality via Stein's method. At the very end of the section we prove Lemma 1, Theorem 2 and sketch the proof of Theorem 1.

Notation. The adjacency relation between vertices v_i and v_j is denoted $v_i \sim v_j$. The indicator of an event A is denoted $\mathbb{I}_A$. In particular, we have

$$\vee_{ijk} = \mathbb{I}_{\{v_i \sim v_j\}}\mathbb{I}_{\{v_j \sim v_k\}}, \qquad \Delta_{i,j,k} = \mathbb{I}_{\{v_i \sim v_j\}}\mathbb{I}_{\{v_j \sim v_k\}}\mathbb{I}_{\{v_k \sim v_i\}}.$$

Introduce random variables $s_{[j,k]} = |S_j \cap S_k|$ and $s_{[i,j,k]} = |S_i \cap S_j \cap S_k|$ and probabilities

$$\begin{aligned}
p_t &= \mathbf{P}(\Delta_{i,j,k} = 1 | s_{[j,k]} = t), \quad q_t = \mathbf{P}(\vee_{kij} = 1 | s_{[j,k]} = t), \\
\bar{p}_t &= \mathbf{P}(s_{[j,k]} = t), \quad p'_t = \mathbf{P}(s_{[i,j,k]} \geq 1 | s_{[j,k]} = t), \\
p''_t &= \mathbf{P}(\Delta_{i,j,k} = 1 | s_{[i,j,k]} = 0, s_{[j,k]} = t).
\end{aligned}$$

We observe that $p_t = q_t$ for $t \geq 1$. Furthermore, we have for $t \geq 0$ that

$$p_t = p'_t + p''_t(1 - p'_t). \tag{10}$$

We denote $p_e = \mathbf{P}(v_i \sim v_j)$ and observe that $\mathbf{E}(\mathbb{I}_{\{v_i \sim v_j\}} | S_i) = \mathbf{E}\mathbb{I}_{\{v_i \sim v_j\}} = p_e$. In particular, we have $p_\vee = p_e^2$. Indeed,

$$\begin{aligned}
p_\vee &= \mathbf{E}\mathbb{I}_{\{v_i \sim v_j\}}\mathbb{I}_{\{v_j \sim v_k\}} = \mathbf{E}\big(\mathbb{I}_{\{v_i \sim v_j\}}\mathbf{E}(\mathbb{I}_{\{v_j \sim v_k\}} | S_i, S_j)\big) \\
&= \mathbf{E}\big(\mathbb{I}_{\{v_i \sim v_j\}} p_e\big) = p_e^2. \tag{11}
\end{aligned}$$

Hoeffding's Decomposition. Let ψ be a real function defined on 3-tuples of subsets of W, which is symmetric in its arguments. We assume that $\mathbf{E}\psi(S_1, S_2, S_3) = 0$. Hoeffding's decomposition [1,6] expands $T = \sum_{\{i,j,k\}\subset[n]} \psi(S_i, S_j, S_k)$ into a series of uncorrelated U statistics

$$T = \binom{n-1}{2} U_1 + (n-2) U_2 + U_3, \tag{12}$$

$$U_1 = \sum_{i\in[n]} f_i, \qquad f_i = \mathbf{E}(\psi(S_i, S_j, S_k)|S_i),$$

$$U_2 = \sum_{\{i,j\}\subset[n]} g_{i,j}, \qquad g_{i,j} = \mathbf{E}(\psi(S_i, S_j, S_k) - f_i - f_j|S_i, S_j),$$

$$U_3 = \sum_{\{i,j,k\}\subset[n]} h_{i,j,k}, \qquad h_{i,j,k} = \psi(S_i, S_j, S_k) - f_i - f_j - f_k - g_{i,j} - g_{i,k} - g_{j,k}.$$

We note that $g(S_i, S_j) := g_{i,j}$ and $h(S_i, S_j, S_k) := h_{i,j,k}$ are symmetric functions of their arguments S_i, S_j and S_i, S_j, S_k and they have the orthogonality property

$$\mathbf{E}(g(S_i, S_j)|S_i) = 0, \qquad \mathbf{E}(h(S_i, S_j, S_k)|S_i, S_j) = 0. \tag{13}$$

(13) implies in particular that all distinct summands f_i, g_{j_1,j_2}, h_{k_1,k_2,k_3} are uncorrelated whatever the indices $i, j_1, j_2, k_1, k_2, k_3$. A simple consequence of (13) is the variance formula

$$\mathbf{Var}T = \mathbf{E}T^2 = \binom{n-1}{2}^2 n\mathbf{E}f_1^2 + (n-2)^2 \binom{n}{2}\mathbf{E}g_{1,2}^2 + \binom{n}{3}\mathbf{E}h_{1,2,3}^2. \tag{14}$$

We construct decomposition (12) for $T = \bar{N}_\triangle$ and $T = \bar{N}_\vee$ and use subscripts $\triangle$ and $\vee$ to distinguish the respective terms $\psi_\triangle$, $f_{\triangle j}$, $g_{\triangle i,j}$, $h_{\triangle i,j,k}$ and $\psi_\vee$, $f_{\vee j}$, $g_{\vee i,j}$, $h_{\vee i,j,k}$.

Decomposition of $\bar{N}_\triangle$. We put $\psi_\triangle(S_i, S_j, S_k) = \triangle_{i,j,k} - p_\triangle$ and apply (12) to $T = \bar{N}_\triangle$. We shall show that for any j and $k \neq j$

$$f_{\triangle j} \equiv 0, \tag{15}$$

$$g_{\triangle j,k} = \sum_{t=1}^{r} (\mathbb{I}_{\{s_{[j,k]}=t\}} - \bar{p}_t) p_t. \tag{16}$$

To show (15), we observe that, given S_j, the conditional probability of triangle induced by v_i, v_j, v_k (the quantity $\mathbf{E}(\triangle_{i,j,k}|S_j)$) does not depend on S_j. Hence $\mathbf{E}(\triangle_{i,j,k}|S_j) = p_\triangle$ and, consequently, $f_{\triangle j} \equiv 0$. To show (16) we observe that, given the pair (S_j, S_k), the conditional probability of the triangle (the quantity $\mathbf{E}(\triangle_{i,j,k}|S_j, S_k)$) only depends on the number $s_{[j,k]}$. In particular, the following random variables are equal

$$\mathbf{E}(\triangle_{i,j,k}|S_j, S_k) = \mathbf{E}(\triangle_{i,j,k}|s_{[j,k]}) = \sum_{t=1}^{r} \mathbb{I}_{\{s_{[j,k]}=t\}} p_t. \tag{17}$$

Taking the expected values of the left and right sides of (17) we obtain the identity

$$p_\Delta = \mathbf{E}\big(\mathbf{E}(\Delta_{i,j,k}|S_j, S_k)\big) = \sum_{t=1}^{r} \bar{p}_t p_t. \tag{18}$$

From (17), (18) we obtain expression (16) for $g_{\Delta j,k} = \mathbf{E}(\Delta_{i,j,k}|S_j, S_k) - p_\Delta$.

Decomposition of $\bar{N}_\vee$. We put $\psi_\vee(S_i, S_j, S_k) = \vee_{ijk} + \vee_{jki} + \vee_{kij} - 3p_\vee$ and apply (12) to $T = \bar{N}_\vee$. We shall show that for any j and $k \neq j$

$$f_{\vee j} \equiv 0, \tag{19}$$

$$g_{\vee j,k} = 2p_e\big(\mathbb{I}_{\{v_j \sim v_k\}} - p_e\big) + \sum_{t=0}^{r}\big(\mathbb{I}_{\{s_{[j,k]}=t\}} - \bar{p}_t\big)q_t \tag{20}$$

$$= \big(\mathbb{I}_{\{s_{[j,k]}=0\}} - \bar{p}_0\big)q_0 + \sum_{t=1}^{r}\big(\mathbb{I}_{\{s_{[j,k]}=t\}} - \bar{p}_t\big)(q_t + 2p_e). \tag{21}$$

To show (19), we observe that, given S_j, the conditional probabilities of cherries $\vee_{ijk}, \vee_{jki}, \vee_{kij}$ (the random variables $\mathbf{E}(\vee_{ijk}|S_j), \mathbf{E}(\vee_{jki}|S_j), \mathbf{E}(\vee_{kij}|S_j)$) do not depend on S_j. Hence all three conditional probabilities equal to the unconditional one, $p_\vee$. Consequently, $f_{\vee j} \equiv 0$. To show (20) we observe that, given the pair (S_j, S_k), the conditional probability

$$\mathbf{E}\big(\vee_{ijk}|S_j, S_k\big) = \mathbf{E}\big(\mathbb{I}_{\{v_i \sim v_j\}}\mathbb{I}_{\{v_j \sim v_k\}}\big|S_j, S_k\big) = \mathbb{I}_{\{v_j \sim v_k\}}p_e, \tag{22}$$

cf. (11). Similarly, we obtain $\mathbf{E}\big(\vee_{jki}|S_j, S_k\big) = \mathbb{I}_{\{v_j \sim v_k\}}p_e$. Furthermore, we note that the conditional probability $\mathbf{E}(\vee_{kij}|S_j, S_k)$ only depends on the number $s_{[j,k]}$. Hence the following random variables are equal

$$\mathbf{E}(\vee_{kij}|S_j, S_k) = \mathbf{E}(\vee_{kij}|s_{[j,k]}) = \sum_{t=0}^{r}\mathbb{I}_{\{s_{[j,k]}=t\}}q_t. \tag{23}$$

An immediate consequence of (23) are the identities

$$p_\vee = \mathbf{E}\big(\mathbf{E}(\vee_{kij}|S_j, S_k)\big) = \sum_{t=0}^{r}\bar{p}_t q_t,$$

$$\mathbf{E}(\vee_{kij}|S_j, S_k) - p_\vee = \sum_{t=0}^{r}\big(\mathbb{I}_{\{s_{[j,k]}=t\}} - \bar{p}_t\big)q_t. \tag{24}$$

From (22), (24) we obtain (20) for $g_{\vee j,k} = \mathbf{E}\big(\vee_{ijk} + \vee_{jki} + \vee_{kij}\big|S_j, S_k\big) - 3p_\vee$. Invoking in (20) the identities $\mathbb{I}_{\{v_j \sim v_k\}} = \sum_{t=1}^{r}\mathbb{I}_{\{s_{[j,k]}=t\}}$ and $p_e = \sum_{t=1}^{r}\bar{p}_t$ we obtain (21).

Variances. *Variance of N_Δ.* Assuming that $r^3 = O(m)$ we show below that

$$\mathbf{E}g_{\Delta j,k}^2 = \frac{r^4}{m^3} + 2\frac{r^7}{m^4} + \frac{r^{10}}{m^5} + O\Big(\frac{r^6}{m^4}\Big), \tag{25}$$

$$\mathbf{E}h_{\Delta i,j,k}^2 = \frac{r^3}{m^2} + \frac{r^6}{m^3} + O\Big(\frac{r^5}{m^3}\Big). \tag{26}$$

Proof of (25). Using $\mathbb{I}_{\{s_{[j,k]}=t\}}\mathbb{I}_{\{s_{[j,k]}=u\}} = 0$, for $t \neq u$, we obtain from (16) that

$$\mathbf{E}g_{j,k}^2 = \sum_{t=1}^{r} p_t^2 \bar{p}_t(1 - \bar{p}_t) - 2 \sum_{1 \leq s < t \leq r} p_s p_t \bar{p}_s \bar{p}_t. \tag{27}$$

We write $\mathbf{E}g_{\triangle j,k}^2 = p_1^2 \bar{p}_1 + R_1$, where the remainder term

$$R_1 = \sum_{t=2}^{r} p_t^2 \bar{p}_t(1 - \bar{p}_t) - p_1^2 \bar{p}_1^2 - 2 \sum_{1 \leq s < t \leq r} p_s p_t \bar{p}_s \bar{p}_t. \tag{28}$$

Combining (10) and Lemma 3, see also (64), we obtain for $r^3 = O(m)$ that

$$p_1^2 \bar{p}_1 = r^4 m^{-3} + O(r^7 m^{-4}). \tag{29}$$

Finally, (29) together with the bound $R_1 = O(r^6 m^{-4})$ of Lemma 4 imply (25).

Proof of (26). We observe that (13), (15) implies $\mathbf{E}(\psi_\triangle(S_i, S_j, S_k)|S_i, S_k) = g_{\triangle i,k}$. Furthermore, the identity $\mathbf{E}(g_{\triangle i,j}|S_i) = 0$ implies $\mathbf{E}g_{\triangle j,k}g_{\triangle i,h} = 0$, for $\{j, k\} \neq \{i, h\}$. Combining these identities we obtain

$$\mathbf{E}h_{\triangle i,j,k}^2 = \mathbf{E}\psi_\triangle^2(S_i, S_j, S_k) - 3\mathbf{E}g_{\triangle i,j}^2 = p_\triangle(1 - p_\triangle) - 3\mathbf{E}g_{\triangle i,j}^2. \tag{30}$$

For $r^3 = O(m)$ relation (30) combined with (25) and Lemma 4(ii) imply (26).

Variance of $N_\vee$. Assuming that $r^3 = O(m)$ we show below that

$$\mathbf{E}g_{\vee j,k}^2 = (r^4 + 4r^5 + 4r^6)m^{-3} + O(r^8 m^{-4}), \tag{31}$$

$$\mathbf{E}h_{\vee i,j,k}^2 = (3r^4 + 6r^3)m^{-2} + 3r^6 m^{-3} + O(r^5 m^{-3}). \tag{32}$$

Proof of (31). Using $\mathbb{I}_{\{s_{[j,k]}=t\}}\mathbb{I}_{\{s_{[j,k]}=u\}} = 0$, for $t \neq u$, we obtain from (21) that

$$\mathbf{E}g_{\vee j,k}^2 = \bar{p}_0(1 - \bar{p}_0)q_0^2 + \sum_{t=1}^{r} \bar{p}_t(1 - \bar{p}_t)(q_t + 2p_e)^2 \tag{33}$$

$$- 2\sum_{t=1}^{r} \bar{p}_0 \bar{p}_t q_0(q_t + 2p_e) - 2 \sum_{1 \leq s < t \leq r} \bar{p}_s \bar{p}_t(q_s + 2p_e)(q_t + 2p_e).$$

We write $\mathbf{E}g_{\vee j,k}^2 = \bar{p}_1(q_1 + 2p_e)^2 + R_2$, where $\bar{p}_1(q_1 + 2p_e)^2$ is the leading term. The remainder

$$R_2 = \bar{p}_0(1 - \bar{p}_0)q_0^2 - \bar{p}_1^2(q_1 + 2p_e)^2 + \sum_{t=2}^{r} \bar{p}_t(1 - \bar{p}_t)(q_t + 2p_e)^2 \tag{34}$$

$$- 2\sum_{t=1}^{r} \bar{p}_0 \bar{p}_t q_0(q_t + 2p_e) - 2 \sum_{1 \leq s < t \leq r} \bar{p}_s \bar{p}_t(q_s + 2p_e)(q_t + 2p_e).$$

Combining (10) and Lemma 3 we obtain for $r^3 = O(m)$ that

$$\bar{p}_1(q_1 + 2p_e)^2 = (r^4 + 4r^5 + 4r^6)m^{-3} + O(r^8 m^{-4}). \tag{35}$$

Now, (35) together with the bound $R_2 = O(r^8 m^{-4})$ of Lemma 4 imply (31).

Proof of (32). Proceeding similarly as in the proof of (30) above we obtain

$$\mathbf{E}h_{\vee\,i,j,k}^2 = \mathbf{E}\psi_\vee^2(S_i, S_j, S_k) - 3\mathbf{E}g_{\vee\,i,j}^2. \tag{36}$$

We evaluate the first term on the right

$$\mathbf{E}\,\psi_\vee^2(S_i, S_j, S_k) = \mathbf{Var}\,\psi_\vee(S_i, S_j, S_k) = \mathbf{E}(\vee_{ijk} + \vee_{jki} + \vee_{kij})^2 - (3p_\vee)^2$$
$$= 3p_\vee + 6p_\Delta - 9p_\vee^2 = (3r^4 + 6r^3)m^{-2} + 3r^6m^{-3} + O(r^5m^{-3}). \tag{37}$$

In the first step of (37) we used the identities $\vee_{ijk}\vee_{jki} = \Delta_{i,j,k}$ and $\mathbf{E}(\vee_{ijk})^2 = \mathbf{E}\vee_{ijk} = p_\vee$. In the second step we used the identity $p_\vee = p_e^2$ and Lemma 4(ii), (iii). Finally, (36) combined with (31) and (37) yield (32).

Covariance of N_Δ and $N_\vee$. By the orthogonality property (13) and symmetry,

$$\mathbf{E}\bar{N}_\Delta\bar{N}_\vee = (n-2)^2\binom{n}{2}\mathbf{E}(g_{\Delta\,1,2}g_{\vee\,1,2}) + \binom{n}{3}\mathbf{E}(h_{\Delta\,1,2,3}h_{\vee\,1,2,3}). \tag{38}$$

Assuming that $r^3 = O(m)$ we show below that

$$\mathbf{E}(g_{\Delta\,1,2}g_{\vee\,1,2}) = \frac{r^4 + 2r^5}{m^3} + \frac{2r^8}{m^4} + O\left(\frac{r^7}{m^4}\right), \tag{39}$$

$$\mathbf{E}(h_{\Delta\,1,2,3}h_{\vee\,1,2,3}) = 3\frac{r^3}{m^2} + 3\frac{r^6}{m^3} + O\left(\frac{r^5}{m^3}\right). \tag{40}$$

Proof of (39). Using $\mathbb{I}_{\{s_{[j,k]}=t\}}\mathbb{I}_{\{s_{[j,k]}=u\}} = 0$, for $t \neq u$, we obtain from (16), (21) that

$$\mathbf{E}(g_{\Delta\,1,2}g_{\vee\,1,2}) = \bar{p}_1(1 - \bar{p}_1)p_1(q_1 + 2p_e) - R_{3.1} + R_{3.2}, \tag{41}$$

where the remainder terms

$$R_{3.1} = \sum_{t=1}^r \bar{p}_t\bar{p}_0p_tq_0, \quad R_{3.2} = \sum_{t=2}^r \bar{p}_t(1-\bar{p}_t)p_t(q_t+2p_e) - \sum_{1\leq t\neq u\leq r} \bar{p}_t\bar{p}_up_t(q_u+2p_e). \tag{42}$$

They are bounded in Lemma 4, $R_{3.i} = O(r^7m^{-4})$, $i = 1, 2$. The leading term

$$\bar{p}_1(1 - \bar{p}_1)p_1(q_1 + 2p_e) = (r^4 + 2r^5)m^{-3} + 2r^8m^{-4} + O(r^7m^{-4}). \tag{43}$$

Here we used (63) and (64), the identity $p_1 = q_1$ and Lemma 4(iii).

Proof of (40). By the orthogonality property (13),

$$\mathbf{E}(h_{\Delta\,1,2,3}h_{\vee\,1,2,3}) = \mathbf{E}(\psi_\Delta(S_1, S_2, S_3)\psi_\vee(S_1, S_2, S_3))$$
$$+ \mathbf{E}(q_{\Delta\,1,2} + q_{\Delta\,1,3} + q_{\Delta\,2,3})(q_{\vee\,1,2} + q_{\vee\,1,3} + q_{\vee\,2,3})$$
$$= 3p_\Delta(1 - p_\vee) + 3\mathbf{E}g_{\Delta\,1,2}g_{\vee\,1,2}. \tag{44}$$

In the last step we used the identities

$$\mathbf{E}\psi_{\Delta1,2,3}\psi_{\vee\,1,2,3} = 3\mathbf{E}(\Delta_{1,2,3} - p_\Delta)(\vee_{123} - p_\vee) = 3p_\Delta(1 - p_\vee),$$

which follow from $\Delta_{1,2,3}\vee_{312} = \Delta_{1,2,3}$. Furthermore, note that Lemma 4(iii), (iv) imply

$$p_\Delta(1 - p_\vee) = r^3m^{-2} + r^6m^{-3} + O(r^5m^{-3}).$$

Invoking this expression and (39) in (44) we obtain (40).

Asymptotic Normality. Let $\{(X_{1n}, X_{2n})\}$ be a sequence of bivariate random vectors and let (η_1, η_2) be a Gaussian vector with $\mathbf{E}\eta_j = 0$, $\mathbf{E}\eta_j^2 = \sigma_j^2$, $j = 1, 2$, and $\mathbf{E}\eta_1\eta_2 = \sigma_{12}$. Recall that the sequence $\{(X_{1n}, X_{2n})\}$ converges in distribution to (η_1, η_2) whenever for any reals a, b we have

$$\mathbf{E}e^{iaX_{1n}+ibX_{2n}} \to \mathbf{E}e^{ia\eta_1+ib\eta_2} \qquad \text{as} \qquad n \to +\infty. \tag{45}$$

Here and below "bold italic" i denotes the imaginary unit. Note that (45) holds if $X_n := aX_{1n} + bX_{2n}$ converges in distribution to the Gaussian random variable $\eta := a\eta_1 + b\eta_2$. Here $\mathbf{E}\eta = 0$ and $\mathbf{E}\eta^2 = a^2\sigma_1^2 + b^2\sigma_2^2 + 2ab\sigma_{12} =: \sigma_*^2$. In order to verify (45) one can use the following sufficient condition for the convergence in distribution of X_n to η, see [10, 11],

$$\frac{d}{dt}\mathbf{E}e^{itX_n} + \sigma_*^2 t\mathbf{E}e^{itX_n} \to 0. \tag{46}$$

We remark that condition (46) refers to the Stein method, [10].

Proofs of Lemma 1 and Theorems 1, 2.

Proof of Lemma 1. (4) and (5) follow from Lemma 4(ii), (iii). (6) and (7) follow from (14) combined with (25), (26) and (31), (32) respectively.

Proof of Theorem 1. Given a, b, we verify (46) for $X_n = a\sigma_\Delta^{-1}\bar{N}_\Delta + b\sigma_\vee^{-1}\bar{N}_\vee$ and $\sigma_*^2 = \mathbf{E}X_n^2$.

In the first step we expand X_n in the series of uncorrelated U-statistics. Invoking expansions (12) of $\bar{N}_\Delta$ and $\bar{N}_\vee$ constructed above we obtain the expansion for X_n,

$$X_n = \sum_{\{i,j\}\subset[n]} g_{i,j} + \sum_{\{i,j,k\}\subset[n]} h_{i,j,k}, \tag{47}$$

$$g_{i,j} = (n-2)\left(a\sigma_\Delta^{-1}g_{\Delta i,j} + b\sigma_\vee^{-1}g_{\vee i,j}\right), \qquad h_{i,j,k} = a\sigma_\Delta^{-1}h_{\Delta i,j,k} + b\sigma_\vee^{-1}h_{\vee i,j,k}.$$

Note that $g(S_i, S_j) = g_{i,j}$ and $h(S_i, S_j, S_k) = h_{i,j,k}$ possess the orthogonality property (13).

In the second step we verify (46). Let $\mathbf{E}_{i_1,\ldots,i_k}$ denote the conditional expectation given all the random variables but $S_{i_1},\ldots,S_{i_k}$. Note that (13), (47) imply

$$\mathbf{E}_{i_1,\ldots,i_k}X_n = \sum_{\{i,j\}\subset[n]\backslash\{i_1,\ldots,i_k\}} g_{i,j} + \sum_{\{i,j,k\}\subset[n]\backslash\{i_1,\ldots,i_k\}} h_{i,j,k}.$$

Denote $X_n^{\{i_1,\ldots,i_k\}} = X_n - \mathbf{E}_{i_1,\ldots,i_k}X_n$ so that $X_n = \mathbf{E}_{i_1,\ldots,i_k}X_n + X_n^{\{i_1,\ldots,i_k\}}$. We split

$$\frac{d}{dt}e^{itX_n} = \mathbf{E}iX_n e^{itX_n} = \binom{n}{2}\mathbf{E}ig_{1,2}e^{itX_n} + \binom{n}{3}\mathbf{E}ih_{1,2,3}e^{itX_n} \tag{48}$$

and expand the exponents on the right

$$e^{itX_n} = e^{it\mathbf{E}_{1,2}X_n + itX_n^{\{1,2\}}} = e^{it\mathbf{E}_{1,2}X_n}\left(1 + itX_n^{\{1,2\}} + R_1\right), \qquad (49)$$

$$e^{itX_n} = e^{it\mathbf{E}_{1,2,3}X_n + itX_n^{\{1,2,3\}}} = e^{it\mathbf{E}_{1,2,3}X_n}\left(1 + itX_n^{\{1,2,3\}} + R_2\right). \qquad (50)$$

Invoking these expressions in (48) we obtain (with $o(1)$ accounting for remainders R_1, R_2)

$$\frac{d}{dt}e^{itX_n} = \binom{n}{2}\mathbf{E}i^2 tg_{1,2}^2 e^{it\mathbf{E}_{1,2}X_n} + \binom{n}{3}\mathbf{E}i^2 th_{1,2,3}^2 e^{it\mathbf{E}_{1,2,3}X_n} + o(1)$$

$$= -t\binom{n}{2}\mathbf{E}g_{1,2}^2\mathbf{E}e^{it\mathbf{E}_{1,2}X_n} - t\binom{n}{3}\mathbf{E}h_{1,2,3}^2\mathbf{E}e^{it\mathbf{E}_{1,2,3}X_n} + o(1). \qquad (51)$$

Finally, we replace $\mathbf{E}e^{it\mathbf{E}_{1,2}X_n}$ and $\mathbf{E}e^{it\mathbf{E}_{1,2,3}X_n}$ in (51) by $\mathbf{E}e^{itX_n}$ proceeding similarly as in (49), (50) above, and observe that now the right side of (51) reduces to $-\sigma_*^2 t\mathbf{E}e^{itX_n} + o(1)$. We have arrived to (46).

Proof of Theorem 2. Using the notation $N_\Delta^* = \bar{N}_\Delta/\mathbf{E}N_\Delta$, $N_V^* = \bar{N}_V/\mathbf{E}N_V$ and $\mu = \mathbf{E}N_\Delta/\mathbf{E}N_V$ we write $C_\mathcal{G} - 3\mu$ in the form

$$C_\mathcal{G} - 3\mu = 3\mu\left(\frac{N_\Delta^* + 1}{N_V^* + 1} - 1\right) = 3\mu(N_\Delta^* - N_V^*) + R, \qquad (52)$$

where

$$R = 3\mu\left(-N_\Delta^* N_V^* + (1 + N_\Delta^*)\left(\frac{(N_V^*)^2}{1 + N_V^*}\right)\right)$$

is a negligible reminder. Note that $\mathbf{E}\big(3\mu(N_\Delta^* - N_V^*)\big) = 0$ and $\mathbf{E}\big(3\mu(N_\Delta^* - N_V^*)\big)^2 = \sigma^2$. Furthermore, the asymptotic normality of $\big(\sigma_\Delta^{-1}\bar{N}_\Delta, \sigma_V^{-1}\bar{N}_V\big)$, see Theorem 1, implies the asymptotic normality of $3\mu(N_\Delta^* - N_V^*)/\sigma$.

3 Auxiliary Results

Lemma 2. *(See, e.g., [3]). Given integers $1 \le s \le d_1 \le d_2 \le m$, let D_1, D_2 be independent random subsets of the set $W = \{1, \ldots, m\}$ such that D_1 (respectively D_2) is uniformly distributed in the class of subsets of W of size d_1 (respectively d_2). The probabilities $\mathring{p} := \mathbf{P}(|D_1 \cap D_2| = s)$ and $\tilde{p} := \mathbf{P}(|D_1 \cap D_2| \ge s)$ satisfy*

$$\left(1 - \frac{(d_1 - s)(d_2 - s)}{m + 1 - d_1}\right)\binom{d_1}{s}\binom{d_2}{s}\binom{m}{s}^{-1} \le \mathring{p} \le \tilde{p} \le \binom{d_1}{s}\binom{d_2}{s}\binom{m}{s}^{-1}. \qquad (53)$$

Lemma 3. *Let* $m \to +\infty$. *Assume that* $r^2 = o(m)$. *For* $1 \le t \le r$ *we have*

$$\binom{r}{t}^2 \binom{m}{t}^{-1} \left(1 - 2\frac{(r-t)^2}{m}\right) \le \bar{p}_t \le \binom{r}{t}^2 \binom{m}{t}^{-1}, \tag{54}$$

$$\frac{rt}{m}\left(1 - 2\frac{rt}{m}\right) \le p'_t \le \frac{rt}{m}, \tag{55}$$

$$p''_t \le \binom{2r-2t}{2}\binom{r}{2}\binom{m-t}{2}^{-1}, \tag{56}$$

$$p''_t = (r-t)^2 \binom{r}{2}\binom{m-t}{2}^{-1} + O\left(\frac{(r-t)^3 r^3}{m^3}\right), \tag{57}$$

$$q_0 = r^2 \binom{r}{2}\binom{m}{2}^{-1} + O\left(\frac{r^6}{m^3}\right). \tag{58}$$

Proof of Lemma 3. (54) follows from (53). In the proof of (55), (56), (57), (58) we fix S_j, S_k satisfying $s_{[j,k]} = t$. To show (55) we apply (53) to the probability that random set S_i intersects with a given set $S_j \cap S_k$ of size t. To show (56) we note that S_i is a subset of $W' = W \setminus (S_j \cap S_k)$, by the condition $s_{[i,j,k]} = 0$. The event $\Delta_{i,j,k}$ implies that random set S_i intersects with $S'_j = S_j \cap W'$ and with $S'_k = S_k \cap W'$. Therefore S_i has at least two common elements with the set $S'_j \cup S'_k$ of size $2r - 2t$. We bound the probability of the latter event by (53) and obtain (56).

To show (57) we color elements of S_j white and those of S_k black. The event $\Delta_{i,j,k}$ occurs if: (a) random set S_i intersects with given set $S'_j \cup S'_k$ in at least two elements; (b) the intersection contains white and black elements. Furthermore, we split the event (a) into two events: (a_1) S_i and $S'_j \cup S'_k$ have exactly two common elements; (a_2) S_i and $S'_j \cup S'_k$ have three or more common elements. Note that, by (53), the probability of (a_2) is of order $O\left((2r - 2t)^3 r^3 m^{-3}\right)$. We neglect this event and only consider the two step experiment: we firstly check whether (a_1) occurs, and if so, we secondly check whether (b) occurs. By (53), the probability of (a_1) is

$$\binom{2r-2t}{2}\binom{r}{2}\binom{m-t}{2}^{-1} + O\left((2r-2t)^3 r^3 m^{-3}\right)$$

and the probability of (b) given (a_1) is $(r-t)^2 \binom{2r-2t}{2}^{-1}$. Hence (57).

The proof of (58) is similar to that of (57). The event $\vee_{j,i,k}$ occurs if: (a) random set S_i intersects with given set $S_j \cup S_k$ in at least two elements; (b) the intersection contains white and black elements. We split the event (a) into two events: (a_1) S_i and $S_j \cup S_k$ have exactly two common elements; (a_2) S_i and $S_j \cup S_k$ have three or more common elements. Note that, by (53), the probability of (a_2) is of order $O(r^6 m^{-3})$. We neglect this event and only consider the two step experiment: we firstly check whether (a_1) occurs, and if so, we secondly check whether (b) occurs. By (53), the probability of (a_1) is

$$\binom{2r}{2}\binom{r}{2}\binom{m}{2}^{-1} + O(r^6 m^{-3})$$

and the probability of (b) given (a_1) is $r^2\binom{2r}{2}^{-1}$. Hence (58).

Lemma 4. *Let $m \to +\infty$. Assume that $r^3 = O(m)$.*

(i) For R_1 and R_2 defined in (28) and (34) we have $R_1 = O(r^6m^{-4})$ and $R_2 = O(r^8m^{-4})$. For $R_{3.1}$ and $R_{3.2}$ defined in (42) we have $R_{3.1} = O(r^9m^{-5})$, $i = 1, 2$.

(ii) $p_\Delta = r^3m^{-2} + r^6m^{-3} + O(r^5m^{-3})$.

(iii) $p_e = r^2m^{-1} - 0.5r^2(r-1)^2m^{-2} + O(r^6m^{-3})$.

Remark. Proceeding similarly as in the proof of Lemma 4(ii) one can show that $p_\Delta = O(r^3m^{-2} + r^6m^{-3})$ for $r^2 = o(m)$.

Proof of Lemma 4. Let $c > 0$ and assume that $r^3 \le cm$ uniformly in m. We denote by c' a constant which may depend on c, but it is independent of m and r. c' may attain different values in different places.

We begin with showing two auxiliary inequalities. From (54), (55), (56) we have for $t \ge 1$

$$p_t \le p_t' + p_t'' \le rtm^{-1} + 2r^2(r-t)^2(m-t)^{-2} \le (t+2c)rm^{-1} \le c'trm^{-1}, \tag{59}$$

$$\bar{p}_t = (r)_t(r)_t/(t!(m)_t) \le (r^2/m)^t/t!, \tag{60}$$

$$q_t + 2p_e = p_t + 2p_e \le c'r^2m^{-1}. \tag{61}$$

Proof of (i). To show the bound for R_1 we estimate various terms in (28) using (59), (60),

$$\sum_{t=2}^{r} p_t^2 \bar{p}_t(1 - \bar{p}_t) \le \sum_{t=2}^{r} p_t^2 \bar{p}_t \le \sum_{t=2}^{r} c' \frac{(tr)^2}{m^2} \frac{r^4}{t!m^2} \le c' \frac{r^6}{m^4},$$

$$\sum_{s=1}^{r-1} p_s \bar{p}_s \sum_{t=s+1}^{r} p_t \bar{p}_t \le c' \sum_{s=1}^{r-1} \frac{sr}{m} \frac{r^2}{s!m} \sum_{t=s+1}^{r} \frac{tr}{m} \frac{r^4}{t!m^2} \le c' \frac{r^8}{m^5} \sum_{s=1}^{r-1} \frac{s}{s!} \sum_{t=s+1}^{r} \frac{t}{t!} \le c' \frac{r^8}{m^5},$$

$$p_1^2 \bar{p}_1^2 \le r^6m^{-4}.$$

To show the bound for R_2 we estimate various terms in (34) using (1), (58), (59), (60),

$$\sum_{t=2}^{r} \bar{p}_t(1 - \bar{p}_t)(q_t + 2p_e)^2 \le \sum_{t=2}^{r} \bar{p}_t(q_t + 2p_e)^2 \le c' \sum_{t=2}^{m} \frac{r^4}{m^2t!} \frac{r^4}{m^2} \le c' \frac{r^8}{m^4},$$

$$\sum_{t=2}^{r} \bar{p}_0 \bar{p}_t q_0(q_t + 2p_e) \le \sum_{t=2}^{r} \bar{p}_t q_0 \le c' \sum_{t=2}^{r} \frac{r^4}{m^2t!} \frac{r^4}{m^2} \le c' \frac{r^8}{m^4},$$

$$\sum_{s=1}^{r-1} \bar{p}_s(q_s + 2p_e) \sum_{t=s+1}^{r} \bar{p}_t(q_t + 2p_e) \le c' \sum_{s=1}^{r-1} \frac{r^2}{s!m} \frac{r^2}{m} \sum_{t=s+1}^{r} \frac{r^4}{t!m^2} \frac{r^2}{m} \le c' \frac{r^{10}}{m^5},$$

$$\bar{p}_0(1 - \bar{p}_0)q_0^2 \le q_0^2 \le c'r^8m^{-4},$$

$$\bar{p}_1^2(q_1 + 2p_e)^2 \le c'r^8m^{-4}. \tag{62}$$

To show the bound for $R_{3.1}$ we estimate $R_{3.1} \le q_0 \sum_{1 \le t \le r} \bar{p}_1 p_t = O(r^9 m^{-5})$ using upper bounds (62) and (63) for q_0 and $\sum_{1 \le t \le r} \bar{p}_1 p_t$.

To show the bound for $R_{3.2}$ we estimate

$$|R_{3.2}| \le \bar{p}_1 \bar{p}_2 p_1 (q_2 + 2p_e) + \bar{p}_2 \bar{p}_1 p_2 (q_1 + 2p_e) + \sum_{t=2}^{r} \bar{p}_t p_t \sum_{u=2}^{r} \bar{p}_u (q_u + 2p_e).$$

Using $q_t = p_t$ and (1), (59), (60) we bound the first two summands from above by $O(r^9 m^{-5})$ and bound the sum $\sum_{u=2}^{r} \bar{p}_u (q_u + 2p_e) = O(r^6 m^{-3})$. Finally, we apply (63) to $\sum_{t=2}^{r} \bar{p}_t p_t$.

To show (ii) we evaluate/estimate various terms in (18),

$$\sum_{t=2}^{r} p_t \bar{p}_t \le c' \sum_{t=2}^{r} \frac{tr}{m} \frac{r^4}{m^2 t!} \le \frac{r^5}{m^3} \sum_{t=2}^{r} \frac{t}{t!} \le c' \frac{r^5}{m^3},$$

$$p_1 \bar{p}_1 = \frac{r^3}{m^2} + \frac{r^6}{m^3} + O\left(\frac{r^5}{m^3}\right). \tag{63}$$

In the first line we applied (59), (60). In the second line we invoked the relations

$$\bar{p}_1 = \frac{r^2}{m} + O\left(\frac{r^4}{m^2}\right), \qquad p_1 = \frac{r}{m} + \frac{r^4}{m^2} + O\left(\frac{r^3}{m^2}\right), \tag{64}$$

which follow from (54), (55), (56).

To show (iii) we split $p_e = \bar{p}_1 + \mathbf{P}(s_{[j,k]} \ge 2)$ and evaluate

$$\mathbf{P}(s_{[j,k]} \ge 2) = 0.5 r^2 (r-1)^2 m^{-2} + O(r^6 m^{-3}), \tag{65}$$

$$\bar{p}_1 = r \binom{m-r}{r-1} \binom{m}{r}^{-1} = \frac{r^2}{m} \frac{(m-r)_{r-1}}{(m-1)_{r-1}} = \frac{r^2}{m} - \frac{r^2 (r-1)^2}{m^2} + O\left(\frac{r^6}{m^3}\right). \tag{66}$$

In (65) we applied Lemma 2. In (66) we evaluated the fraction

$$\frac{(m-r)_{r-1}}{(m-1)_{r-1}} = 1 - \frac{(r-1)^2}{m} + O\left(\frac{r^4}{m^2}\right)$$

using the relations

$$\ln\left(\frac{(m-i-1)_{r-1}}{m^t}\right) = \sum_{j=1}^{r-1} \ln\left(1 - \frac{j+i}{m}\right) = -\sum_{j=1}^{r-1} \frac{j+i}{m} + O\left(\frac{r^3}{m^2}\right)$$

$$= -i\frac{r-1}{m} - \sum_{j=1}^{r-1} \frac{j}{m} + O\left(\frac{r^3}{m^2}\right)$$

for $i = 0$ and $i = r - 1$.

References

1. Bentkus, V., Götze, F., van Zwet, W.R.: An edgeworth expansion for symmetric statistics. Ann. Stat. **25**, 851–896 (1997)
2. Bloznelis, M.: On combinatorial Hoeffding decomposition and asymptotic normality of sub-graph count statistics. In: Drmota, M., Flajolet, P., Gardy, D., Gittenberger, B. (eds.) Mathematics and Computer Science III. Algorithms, Trees, Combinatorics and Probabilities, pp. 73–79. Trends in Mathematics, Birkhauser (2004)
3. Bloznelis, M.: Degree and clustering coefficient in sparse random intersection graphs. Ann. Appl. Probab. **23**, 1254–1289 (2013)
4. Bloznelis, M., Godehardt, E., Jaworski, J., Kurauskas, V., Rybarczyk, K.: Recent progress in complex network analysis: models of random intersection graphs. In: Lausen, B., Krolak-Schwerdt, S., Böhmer, M. (eds.) Data Science, Learning by Latent Structures, and Knowledge Discovery. SCDAKO, pp. 69–78. Springer, Heidelberg (2015). https://doi.org/10.1007/978-3-662-44983-7_6
5. Eschenauer, L., Gligor, V.D.: A key-management scheme for distributed sensor networks. In: Proceedings of the 9th ACM Conference on Computer and Communications Security, pp. 41–47 (2002)
6. Hoeffding, V.: A class of statistics with asymptotically normal distribution. Ann. Math. Stat. **19**, 293–325 (1948)
7. Janson, S., Nowicki, K.: The asymptotic distributions of generalized U-statistics with applications to random graphs. Probab. Theory Relat. Fields **90**, 341–375 (1991)
8. Newman, M.E.J., Strogatz, S.H., Watts, D.J.: Random graphs with arbitrary degree distributions and their applications. Phys. Rev. E **64**, 026118 (2002)
9. Rybarczyk, K., Stark, D.: Poisson approximation of the number of cliques in a random intersection graph. J. Appl. Probab. **47**(3), 826–840 (2010)
10. Stein, C.: A bound for the error in the normal approximation to the distribution of a sum of dependent random variables. In: Proceedings of the Sixth Berkeley Symposium on Mathematical Statistics and Probability, vol. 2, pp. 583–602 (1970)
11. Tikhomirov, A.N.: On the convergence rate in the central limit theorem for weakly dependent random variables. Theory Probab. Appl. **25**, 790–809 (1981)
12. Wasserman, S., Faust, K.: Social Network Analysis: Methods and Applications. Cambridge University Press, Cambridge (1994)
13. Yagan, O., Makowski, A.M.: Random key graphs can they be small worlds? In: 2009 First International Conference on Networks and Communications, pp. 313–318 (2009)
14. Zhao, J., Yagan, O., Gligor, V.: Random intersection graphs and their applications in security, wireless communication, and social networks (2015). arXiv:1504.03161

Clustering Properties of Spatial Preferential Attachment Model

Lenar Iskhakov[1], Bogumił Kamiński[2], Maksim Mironov[1], Paweł Prałat[4(✉)], and Liudmila Prokhorenkova[1,3]

[1] Advanced Combinatorics and Network Applications Lab,
Moscow Institute of Physics and Technology, Moscow, Russia
[2] Warsaw School of Economics, Warsaw, Poland
[3] Machine Intelligence and Research Department, Yandex, Moscow, Russia
ostroumova-la@yandex.ru
[4] Department of Mathematics, Ryerson University, Toronto, Canada
pralat@ryerson.ca

Abstract. In this paper, we study the clustering properties of the Spatial Preferential Attachment (SPA) model introduced by Aiello et al. in 2009. This model naturally combines geometry and preferential attachment using the notion of spheres of influence. It was previously shown in several research papers that graphs generated by the SPA model are similar to real-world networks in many aspects. For example, the vertex degree distribution was shown to follow a power law. In the current paper, we study the behaviour of $C(d)$, which is the average local clustering coefficient for the vertices of degree d. This characteristic was not previously analyzed in the SPA model. However, it was empirically shown that in real-world networks $C(d)$ usually decreases as d^{-a} for some $a > 0$ and it was often observed that $a = 1$. We prove that in the SPA model $C(d)$ decreases as $1/d$. Furthermore, we are also able to prove that not only the average but the individual local clustering coefficient of a vertex v of degree d behaves as $1/d$ if d is large enough. The obtained results are illustrated by numerous experiments with simulated graphs.

1 Introduction

The evolution of complex networks attracted a lot of attention in recent years. Empirical studies of different real-world networks have shown that such networks have some typical properties: small diameter, power-law degree distribution, clustering structure, and others [8,22]. Therefore, numerous random graph models have been proposed to reflect and predict such quantitative and topological aspects of growing real-world networks [4,5].

The most well studied property of complex networks is their vertex degree distribution. For the majority of studied real-world networks, the degree distribution was shown to follow a heavy-tailed distribution [2,11,23]. Another important property of real-world networks is their clustering structure. One way to characterize the presence of clustering structure is to measure the *clustering*

© Springer International Publishing AG, part of Springer Nature 2018
A. Bonato et al. (Eds.): WAW 2018, LNCS 10836, pp. 30–43, 2018.
https://doi.org/10.1007/978-3-319-92871-5_3

coefficient, which is, roughly speaking, the probability that two neighbours of a vertex are connected. There are two well-known formal definitions: the global clustering coefficient and the average local clustering coefficient (see Sect. 3 for details). At some point, it was believed that for many real-world networks both the average local and the global clustering coefficients tend to non-zero limit as the network becomes large; for example, some numerical values can be found in [22]; however, this statement for the global clustering coefficient is questionable and recently some contradicting theoretical results were presented in [24].

In this paper, we mostly focus on the behaviour of $C(d)$, which is the average local clustering coefficient for the vertices of degree d. It was empirically shown that in real-world networks $C(d)$ usually decreases as $d^{-\psi}$ for some $\psi > 0$ [9,20, 27,28]. In particular, for many studied networks, $C(d)$ scales as d^{-1} [26].

We study the clustering properties of the *Spatial Preferential Attachment* (SPA) model introduced in [1]. This model combines geometry and preferential attachment; the formal definition is given in Sect. 2.1. It was previously shown that graphs generated by the SPA model are similar to real-world networks in many aspects. For example, it was proven in [1] that the vertex degree distribution follows a power law. More details on the properties of the SPA model are given in Sect. 2.2. However, the clustering coefficient $C(d)$ was not previously analyzed for this model, although some clustering properties were analyzed for the generalized SPA model proposed in [13]. It is proved in [13,14] that the average local clustering coefficient converges in probability to a strictly positive limit. Also, the global clustering coefficient converges to a nonnegative limit, which is nonzero if and only if the power-law degree distribution has a finite variance.

In this paper, we prove that the local clustering coefficient $C(d)$ decreases as $1/d$ in the SPA model. We also obtain some bounds for the individual local clustering coefficients of vertices. The obtained theoretical results are compared with and illustrated by numerous experiments on simulated graphs. Our theoretical results are asymptotic in nature, so we empirically investigate how the model behaves for finite size graphs and see that the asymptotic predictions are still close to empirical observations even for small graph sizes. Additionally, we demonstrate that some of our theoretical assumptions are probably too pessimistic and the SPA model behaves even more predictable than we have proven. We also propose an efficient algorithm for generating graphs according to the SPA model which runs much faster than the straightforward implementation.

Proofs of all theoretical results stated in this paper can be found in the journal version [12] that focuses exclusively on asymptotic results of the model. On the other hand, this proceeding version also contains results on simulated graphs and so can be viewed as a complement to the journal version.

2 Spatial Preferential Attachment Model

2.1 Definition

This paper focuses on the *Spatial Preferential Attachment* (SPA) model, which was first introduced by [1]. This model combines preferential attachment with geometry by introducing "spheres of influence" whose volume grows with the degree of a vertex. The parameters of the model are the *link probability* $p \in [0,1]$ and two constants A_1, A_2 such that $0 < A_1 < \frac{1}{p}$, $A_2 > 0$. All vertices are placed in the m-dimensional unit hypercube $S = [0,1]^m$ equipped with the torus metric derived from any of the L_k norms, i.e.,

$$d(x,y) = \min \left\{ ||x - y + u||_k : u \in \{-1,0,1\}^m \right\} \quad \forall x, y \in S.$$

The SPA model generates a sequence of random directed graphs $\{G_t\}$, where $G_t = (V_t, E_t)$, $V_t \subseteq S$. Let $\deg^-(v,t)$ be the in-degree of the vertex v in G_t, and $\deg^+(v,t)$ its out-degree. Then, the *sphere of influence* $S(v,t)$ of the vertex v at time $t \geq 1$ is the ball centered at v with the following volume:

$$|S(v,t)| = \min \left\{ \frac{A_1 \deg^-(v,t) + A_2}{t}, 1 \right\}.$$

In order to construct a sequence of graphs we start at $t = 0$ with G_0 being the null graph. At each time step t we construct G_t from G_{t-1} by, first, choosing a new vertex v_t *uniformly at random* from S and adding it to V_{t-1} to create V_t. Then, independently, for each vertex $u \in V_{t-1}$ such that $v_t \in S(u, t-1)$, a directed link (v_t, u) is created with probability p. Thus, the probability that a link (v_t, u) is added in time-step t equals $p |S(u, t-1)|$.

2.2   Properties of the Model

In this section, we briefly discuss previous studies on properties and applications of the SPA model. This model is known to produce scale-free networks, which exhibit many of the characteristics of real-life networks [1,7]. Specifically, Theorem 1.1 in [1] proves that the SPA model generates graphs with a power-law in-degree distribution with coefficient $1 + 1/(pA_1)$. On the other hand, the average out-degree is asymptotic to $pA_2/(1 - pA_1)$, as shown in Theorem 1.3 in [1]. In [15], it was demonstrated that the SPA model give the best fit, in terms of graph structure, for a series of social networks derived from Facebook. In [16], some properties of common neighbours were used to explore the underlying geometry of the SPA model and quantify vertex similarity based on the distance in the space. Usually, the distribution of vertices in S is assumed to be uniform [16], but [17] also investigated non-uniform distributions, which is clearly a more realistic setting. The SPA model was also used to study a duopoly market on which there is uncertainty of a product quality [18]. Finally, in [25] modularity of this model was investigated, which is a global criterion to define communities and a way to measure the presence of community structure in a network.

3   Clustering Coefficient

Clustering coefficient measures how likely two neighbours of a vertex are connected by an edge. There are several definitions of clustering coefficient proposed in the literature (see, e.g., [5]). The *global clustering coefficient* $C_{glob}(G)$ of a graph G is the ratio of three times the number of triangles to the number of pairs of adjacent edges in G. In other worlds, if we sample a random pair of adjacent vertices in G, then $C_{glob}(G)$ is the probability that these three vertices form a triangle. The global clustering coefficient in the SPA model was previously studied in [13,14] and it was proven that $C_{glob}(G_n)$ converges to a limit, which is positive if and only if the power-law degree distribution has a finite variance.

In this paper, we focus on the *local clustering coefficient*, which was not previously analyzed for the SPA model. Let us first define it for an undirected graph $G = (V, E)$. Let $N(v)$ be the set of neighbours of a vertex v, $|N(v)| = \deg(v)$. For any $B \subseteq V$, let $E(B)$ be the set of edges in the graph induced by the vertex set B; that is, $E(B) = \{(u, w) \in E : u, w \in B\}$. Finally, *clustering coefficient* of a vertex v is defined as follows:

$$c(v) = \frac{|E(N(v))|}{\binom{\deg(v)}{2}}.$$

Clearly, $0 \leq c(v) \leq 1$.

Note that the local clustering $c(v)$ is defined individually for each vertex and it can be noisy, especially for the vertices of not too large degrees. Therefore, the following characteristic was extensively studied in the literature for various real-world networks and some random graph models. Let $C(d)$ be the local clustering coefficient averaged over the vertices of degree d; that is,

$$C(d) = \frac{\sum_{v:\deg(v)=d} c(v)}{|\{v : \deg(v) = d\}|}.$$

Further in the paper we will also use the notation $c(v, t)$ and $C(d, t)$ specifying that the graph has t vertices.

The local clustering $C(d)$ was extensively studied both theoretically and empirically. For example, it was observed in a series of papers that in real-world networks $C(d) \sim d^{-\varphi}$ for some $\varphi > 0$. In particular, [26] shows that $C(d)$ can be well approximated by d^{-1} for four large networks, [28] obtains power-law in a real network with parameter 0.75, while [9] obtains $\varphi = 0.33$. The local clustering coefficient was also studied in several random graph models of complex networks. For instance, it was shown in [10,19,21] that some models have $C(d) \sim d^{-1}$. As we prove in this paper, similar behaviour is also observed in the SPA model.

Recall that the graph G_t constructed according to the SPA model is directed. Therefore, we first analyze the directed version of the local clustering coefficient and then, as a corollary, we obtain the corresponding results for the undirected version. Let us now define the directed clustering coefficient. By $N^-(v, t) \subseteq V_t$ we denote the set of in-neighbours of a vertex v at time t. So, the directed clustering coefficient of a vertex v at time t and the average directed clustering for the vertices of incoming degree d are defined as

$$c^-(v,t) = \frac{|E(N^-(v,t))|}{\binom{\deg^-(v,t)}{2}}, \quad C^-(d,t) = \frac{\sum_{v:\deg^-(v,t)=d} c^-(v,t)}{|\{v : \deg^-(v,t) = d\}|}.$$

Note that we normalize $c^-(v,t)$ by $\binom{\deg^-(v,t)}{2}$, since in the SPA model edges can be created only from younger vertices to older ones.

4   Results

4.1   Notation

Let us start with introducing some notation. As typical in random graph theory, all results in this paper are asymptotic in nature; that is, we aim to investigate properties of G_n for n tending to infinity. We say that an event holds *asymptotically almost surely* (a.a.s.) if it holds with probability tending to one as $n \to \infty$. Also, given a set S we say that *almost all* elements of S have some property P if the number of elements of S that do not have P is $o(|S|)$. We emphasize that the notations $o(\cdot)$ and $O(\cdot)$ refer to functions of n, not necessarily positive, whose growth is bounded. We use the notations $f \ll g$ for $f = o(g)$ and $f \gg g$ for $g = o(f)$. We also write $f(n) \sim g(n)$ if $f(n)/g(n) \to 1$ as $n \to \infty$ (that is, when $f(n) = (1 + o(1))g(n)$). Finally, by $f(n) = \Omega(g(n))$ we denote the fact that f is asymptotically bounded below by g and by $f(n) = \Theta(g(n))$ that f is asymptotically bounded both above and below by g.

First we consider the directed clustering coefficient. It turns out that for the SPA model we are able not only to prove the asymptotics for $C^-(d,n)$, which is the average clustering over all vertices of in-degree d, but also analyze the individual clustering coefficients $c^-(v,n)$. However, in order to do this, we need to assume that $\deg^-(v,n)$ is large enough.

From technical point of view, it will be convenient to partition the set of contributing edges, $E(N^-(v,n))$, and independently consider edges to "old" and to "young" neighbours of v. Formally, for a given function $\omega(n)$ that tends to infinity as $n \to \infty$, let $\hat{T}_v$ be the smallest integer t such that $\deg^-(v,t)$ exceeds $\omega \log n$ (or $\hat{T}_v = n$ if $\deg^-(v,n) < \omega \log n$). Vertices in $N^-(v, \hat{T}_v)$ are called *old neighbours of v*; $N^-(v,n) \setminus N^-(v, \hat{T}_v)$ are *new neighbours of v*. Finally,

$$E_{old}(N^-(v,n)) = \{(u,w) \in E_n : u \in N^-(v,n), w \in N^-(v, \hat{T}_v)\},$$
$$E_{new}(N^-(v,n)) = E(N^-(v,n)) \setminus E_{old}(N^-(v,n));$$

and

$$c^-(v,n) = c_{old}(v,n) + c_{new}(v,n), \tag{1}$$

where

$$c_{old}(v,n) = |E_{old}(N^-(v,n))| \Big/ \binom{\deg^-(v,n)}{2},$$

$$c_{new}(v,n) = |E_{new}(N^-(v,n))| \Big/ \binom{\deg^-(v,n)}{2}.$$

4.2   Results

Let us start with the following theorem which is extensively used in our reasonings and is interesting and important on its own. Variants of this results were proved in [16, 17]; here, we present a slightly modified statement from [17], adjusted to our current needs.

Theorem 1. *Let* $\omega = \omega(n)$ *be any function tending to infinity together with* n. *The following holds with probability* $1 - o(n^{-4})$. *For any vertex* v *with* $\deg^-(v, n) = k = k(n) \geq \omega \log n$ *and for all values of* t *such that*

$$n \left(\frac{\omega \log n}{k} \right)^{\frac{1}{pA_1}} =: T_v \leq t \leq n,$$

we have

$$\deg^-(v, t) \sim k \left(\frac{t}{n} \right)^{pA_1}.$$

The expression for T_v is chosen so that at this time vertex v has a.a.s. $(1 + o(1))\omega \log n$ neighbours. The implication of this theorem is that once a vertex accumulates $\omega \log n$ neighbours, its behaviour can be predicted with high probability until the end of the process (that is, till time n).

Let us note that Theorem 1 immediately implies the following two corollaries.

Corollary 1. *Let* $\omega = \omega(n)$ *be any function tending to infinity together with* n. *The following holds with probability* $1 - o(n^{-4})$. *For every vertex* v, *and for every time* T *so that* $\deg^-(v, T) \geq \omega \log n$, *for all times* t, $T \leq t \leq n$,

$$\deg^-(v, t) \sim \deg^-(v, T) \left(\frac{t}{T} \right)^{pA_1}.$$

Corollary 2. *Let* $\omega = \omega(n)$ *be any function tending to infinity together with* n. *The following holds with probability* $1 - o(n^{-4})$. *For any vertex* v_i *born at time* $i \geq 1$, *and* $i \leq t \leq n$ *we have that* $\deg^-(v_i, t) \leq \omega \log n \, (t/i)^{pA_1}$.

Theorem 1 can be used to show that the contribution to $c^-(v, n)$ coming from edges to new neighbours of v is well concentrated.

Theorem 2. *Let* $\omega = \omega(n)$ *be any function tending to infinity together with* n. *Then, with probability* $1 - o(n^{-1})$ *for any vertex* v *with*

$$\deg^-(v, n) = k = k(n) \geq (\omega \log n)^{4+(4pA_1+2)/(pA_1(1-pA_1))}$$

we have

$$c_{new}(v, n) = \Theta(1/k).$$

Unfortunately, if a vertex v lands in a densely populated region of S, it might happen that $c_{old}(v, n)$ is much larger than $1/k$. We show the following 'negative' result (without trying to aim for the strongest statement) that shows that there is no hope for extending Theorem 2 to $c^-(v, n)$.

Theorem 3. *Let* $C = 5\log(1/p)$ *and* $\xi = \xi(n) = 1/(\omega(\log\log n)^2$ $(\log\log\log n)) = o(1)$ *for some* $\omega = \omega(n)$ *tending to infinity as* $n \to \infty$*. Suppose that* $k = k(n)$ *is such that* $2 \leq k \leq n^{\xi}$*. Then, a.a.s., there exists a vertex* v *such that* $\deg^-(v,n) \sim k$ *and*

(i) $c^-(v,n) = 1$*, provided that* $2 \leq k \leq \sqrt{\log n/C}$*,*
(ii) $c^-(v,n) = \Omega(1) \gg 1/k$*, provided that* $\sqrt{\log n/C} \leq k \leq \log n/\log\log n$*,*
(iii) $c^-(v,n) \gg (\log\log n)^2(\log\log\log n)/k \gg 1/k$*, provided that* $\log n/\log\log n \leq k \leq n^{\xi}$*.*

On the other hand, Theorem 2 implies immediately the following corollary.

Corollary 3. *Let* $\omega = \omega(n)$ *be any function tending to infinity together with* n*. The following holds with probability* $1 - o(n^{-1})$*. For any vertex* v *for which*

$$\deg^-(v,n) = k = k(n) \geq (\omega\log n)^{4+(4pA_1+2)/(pA_1(1-pA_1))}$$

it holds that

$$c^-(v,n) \geq c_{new}(v,n) = \Omega(1/k)$$
$$c^-(v,n) = c_{old}(v,n) + c_{new}(v,n) = O(\omega\log n/k) + O(1/k) = O(\omega\log n/k).$$

Moreover, despite the above 'negative' result, almost all vertices (of large enough degrees) have clustering coefficients of order $1/k$. Here is a precise statement. The conclusions in cases (i)' and (ii)' follow immediately from Theorem 2.

Theorem 4. *Let* $\varepsilon, \delta \in (0, 1/2)$ *be any two constants, and let* $k = k(n) \leq n^{pA_1-\varepsilon}$ *be any function of* n*. Let* X_k *be the set of vertices of* G_n *of in-degree between* $(1-\delta)k$ *and* $(1+\delta)k$*. Then, a.a.s., the following holds.*

(i) Almost all vertices in X_k *have* $c_{old}(v,n) = O(1/k)$*, provided that* $k \gg \log^{C_1} n$*, where* $C_1 = (1 + (2+\varepsilon)pA_1)/(1-pA_1)$*.*
(i)' As a result, almost all vertices in X_k *have* $c^-(v,n) = \Theta(1/k)$*, provided that* $k \gg \log^C n$*, where* $C = 4 + (4pA_1 + 2)/(pA_1(1-pA_1))$*.*
(ii) The average clustering coefficient $c_{old}(v,n)$ *of vertices in* X_k *is* $O(1/k)$*; that is,*

$$\frac{1}{|X_k|}\sum_{v\in X_k} c_{old}(v,n) = O(1/k),$$

provided that $k \gg \log^{C_2} n$*, where* $C_2 = (1 + (2 + pA_1 + \varepsilon)pA_1)/(1-pA_1)$*.*
(ii)' As a result, the average clustering coefficient $c^-(v,n)$ *of vertices in* X_k *is* $\Theta(1/k)$*; that is,*

$$\frac{1}{|X_k|}\sum_{v\in X_k} c^-(v,n) = \Theta(1/k),$$

provided that $k \gg \log^C n$*, where* $C = 4 + (4pA_1 + 2)/(pA_1(1-pA_1))$*.*

Finally, let us briefly discuss the undirected case. The following lemma holds.

Lemma 1. *Let $\omega = \omega(n)$ be any function tending to infinity together with n. The following holds with probability $1 - o(n^{-3})$. For every vertex v_i,*

$$\deg^+(v_i, i) = \deg^+(v_i, n) \leq \omega \log n.$$

Note that a weaker bound of $\log^2 n$ was proved in [1]; with Corollary 2 in hand, we can get slightly better bound but the argument remains the same.

From the above lemma we get the following corollary.

Corollary 4. *Let $c(v, n)$ be the clustering coefficient defined for the undirected graph $\hat{G}_n$ obtained from G_n by considering all edges as undirected. Then Corollary 3 and Theorem 4 hold with replacing $c^-(v, n)$ by $c(v, n)$.*

Indeed, according to Lemma 1, a.a.s. the out-degrees of all vertices do not exceed $\omega \log n$. Therefore, even if out-neighbours of a vertex form a complete graph, the contribution from them is at most $\binom{\omega \log n}{2}$, which is much smaller than the required lower bound for k.

5   Experiments

In this section, we illustrate the theoretical, asymptotic, results presented in the previous section by analyzing the local clustering coefficient for graphs of various orders generated according to the SPA model.

5.1   Algorithm

Let us first discuss the complexity of the straightforward (*naive*) algorithm for generating graphs according to the SPA model. At each step we add one vertex and, for each existing vertex, we check if the new vertex belongs to its sphere of influence. Then we (possibly) add new edges and update the radii for all vertices. The complexity of this procedure is $\Theta(n^2)$.

Let us now propose a more efficient algorithm. First, we describe this algorithm and provide heuristic arguments about its complexity. Then, we compare running times of the new algorithm and the naive one.

Our algorithm works in several phases, as described further in the text. For now, let us assume that we already generated a graph on n vertices according to the SPA model and we want to add one additional vertex. It is known that

$$\mathbb{E}\Big(\deg^-(v_i, t) \Big) \sim \frac{A_2}{A_1} \left(\frac{t}{i}\right)^{pA_1},$$

provided that $i \gg 1$ (see, for example, [7]). We call a vertex *heavy* if its degree is at least D for some D; otherwise, it is *light*. All heavy vertices are kept in a separate list H. Fix

$$D = \frac{A_2}{A_1} \left(\frac{n}{T}\right)^{pA_1}, \tag{2}$$

so H has expected size around T. The choice of an optimal value of T will be discussed further in this section.

Let us divide $S = [0,1]^2$ into k squares where k is some perfect square; that is, each square will have side length $1/\sqrt{k}$. (We choose the dimension $m = 2$ for our simulations. However, the ideas can easily be applied for an arbitrary m.) All light vertices are kept in k disjoint lists; let $L(i)$ be a list containing all light vertices from square i. The expected number of vertices in each list is $(n-T)/k$.

We want the following property to be satisfied:

$$\sqrt{\frac{A_1 D + A_2}{\pi n}} \leq \frac{1}{\sqrt{k}}. \tag{3}$$

Indeed, if this is the case, then no light vertex v_i has the area of influence that touches squares other than the square containing v_i and the 8 adjacent squares. Moreover, the same property will hold for all $t > n$ as areas of influence of light vertices decrease with time. Hence, since we aim for an integer $\sqrt{k}$ to be as large as possible:

$$k = \left\lfloor \sqrt{\frac{\pi n}{A_1 D + A_2}} \right\rfloor^2 \Rightarrow k \approx \frac{\pi n}{A_2 \left(1 + (n/T)^{pA_1}\right)}. \tag{4}$$

The most expensive computational work for the algorithm is the number of comparisons needed in order to add a vertex v_{n+1} to a graph, which is of order

$$f(T) := T + 9\frac{n-T}{k} = T + \frac{9A_2}{\pi}(1 - T/n)\left(1 + (n/T)^{pA_1}\right). \tag{5}$$

Hence, the function $f(T)$ is minimized for

$$T = \frac{9npA_1 A_2(n/T)^{pA_1}}{\pi n - 9A_2 - 9A_2(1 - pA_1)(n/T)^{pA_1}}.$$

For large n the second and the third terms in the denominator are negligible, as $pA_1 < 1$; moreover, if pA_1 is close to 1 we will soon show that $T = \Theta(n^{1/2-\epsilon})$ for some small $\epsilon > 0$, so the approximation converges fast. Thus, we may approximate T by:

$$T \approx n^{1-1/(pA_1+1)} \left(\frac{9pA_1 A_2}{\pi}\right)^{1/(pA_1+1)}. \tag{6}$$

Using this T we can calculate the recommended value of D, see (3), and the density of the $\sqrt{k} \times \sqrt{k}$ grid, see (4).

Below are some practical implementation details:

– It is computationally expensive to recalculate H and L division each time a new vertex is added. By empirical testing, we have found that the recalculation should be done approximately after adding $t/4$ vertices, where t is the number of vertices in already constructed graph. As a result, the number of phases is $O(\log n)$, as each time the number of vertices increases by approximately 25%.

- As we work in phases, at each step we have to check if some light vertex becomes heavy, and move it to the appropriate list, if needed. However, this operation is not expensive computationally.
- After several phases, for actually constructed graphs the optimal parameters k, T and D might deviate from the theoretical values derived above. Therefore, in the implementation we choose the optimal parameters conditional on the actual input graph structure. Namely, for each candidate value k we can calculate the corresponding D using (3) and then calculate T from the data (this is the actual number of heavy vertices given D). We choose k to optimize the number of comparisons needed to add one vertex to the actual graph, the approximation for this value is given in (5). After that we dynamically construct H and L lists.

Let us now discuss the complexity of the obtained algorithm. Equation (5) shows that T is expected to be of order $n^{pA_1/(pA_1+1)}$. So, we may derive from (4) that k is of order $n^{1-pA_1+(pA_1)^2/(pA_1+1)} = n^{1/(pA_1+1)}$. From (5) we obtain that $f(T)$ grows as $n^{pA_1/(pA_1+1)}$. So, the expected complexity of the whole process is $\Theta\left(n^{2-1/(pA_1+1)}\right) \ll n^2$.

Figure 1 presents an empirical comparison of the running time for new and naive algorithms. We also present this figure in log-log scale. The computations were performed using Julia 0.6.2 language [3] and LightGraphs [6] package on a single thread of Intel i5-5200U @ 2.20 GHz processor.

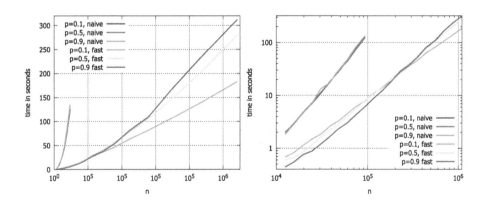

Fig. 1. Running time of the proposed and the naive algorithms.

Finally, let us mention that further improvements of the algorithm are possible. For example, one can keep more than two lists H and L. For example, $L_s(i)$ could contain vertices of degree between 2^{s-1} and 2^s that landed in region i, so the total number of lists is $O(\log n)$. Then, the running time of the algorithm would be $O(n \log n)$. Indeed, during a phase that started at time t, L_s has expected size $O(t\, 2^{-s/pA_1})$; since vertices from $L_s(i)$ are gathered from the square of area, say, $2^s/t$, the expected size of this list is $O(2^{s-s/(pA_1)}) = O(1)$.

Hence, after adding one vertex, $O(\log n)$ lists are investigated and we expect only a constant number of comparisons done on each list. Of course, there is always a trade-off between the running time of an algorithm and how complicated it is to implement it. For our purpose, we decided to go for a simpler algorithm with only two lists.

5.2 Empirical Analysis of the Local Clustering Coefficient

In this section, we compare asymptotic theoretical results obtained in Sect. 4 with empirical results obtained for graphs with finite n. All graphs are generated according to the algorithm described in Sect. 5.1.

It is proven in Theorem 4 that $\frac{1}{X_d} \sum_{v \in X_d} c^-(v, n) = \Theta(1/d)$ for $d \gg \log^C n$, where $C = 4 + (4pA_1 + 2)/(pA_1(1 - pA_1))$. In order to illustrate this result, we generated 10 graphs for each $p \in \{0.1, 0.2, \ldots, 0.9\}$, $A_1 = 1$, $A_2 = 10(1-p)/p$ (A_2 is chosen to fix the expected asymptotic degree equal 10) and computed the average value of $C^-(d, n)$ for $n = 10^6$, see Fig. 2 (left). Similarly, Fig. 2 (right) presents the same measurements for the undirected average local clustering $C(d, n)$. Note that in both cases figures agree with our theoretical results: both $C^-(d, n)$ and $C(d, n)$ decrease as c/d with some c for large enough d (we added a function $10/d$ for comparison). Note that for small p the maximum degree is small, therefore the sizes of the generated graphs are not large enough to observe a straight line in log-log scale.

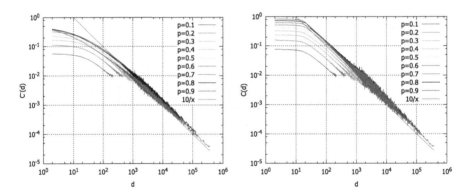

Fig. 2. Average local clustering coefficient for directed (left) and undirected (right) graphs.

Note that for all $p \in (0, 1)$ we have $C = 4 + \frac{4p+2}{p(1-p)} > 18$, so, our theoretical results are expected to hold for $d \gg \log^C n > 10^{20}$ which is irrelevant as the order of the graph is only 10^6. However, we observe the desired behaviour for much smaller values of d; that is, in some sense, our bound is too pessimistic.

Also, note that the statement $C^-(d, n) = \Theta(1/d)$ is stronger that the statement of Theorem 4, since in the theorem we averaged $c^-(v, n)$ over the set X_d

of vertices of in-degree between $(1 - \delta)d$ and $(1 + \delta)d$. In order to illustrate the difference, on Fig. 3 we present the smoothed curves for the directed (left) and undirected (right) local clustering coefficients averaged over X_d for $\delta = 0.1$. Note that this smoothing substantially reduce the noise observed on Fig. 2.

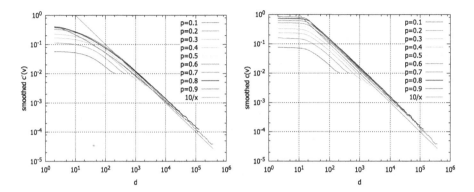

Fig. 3. Local clustering coefficient for directed (left) and undirected (right) graphs averaged over X_d.

Next, let us illustrate the fact that the number of edges between "new" neighbours of a vertex is more predictable than the number of edges going from some neighbours to "old" ones. We extensively used this difference in Sect. 4.2, where we analyzed new and old edges separately. In our experiments, we split $c^-(v, n)$ into "old" and "new" parts as in (1), but now we take $\hat{T}_v$ be the smallest integer t such that $\deg^-(v, t)$ exceeds $\deg^-(v, n)/2$. As a result, we compute the average local clustering coefficients $C_{old}^-(d)$ and $C_{new}^-(d)$. Figure 4 shows that $C_{new}^-(d)$ can almost perfectly be fitted by c/d with some c, while most of the noise comes from $C_{old}^-(d)$.

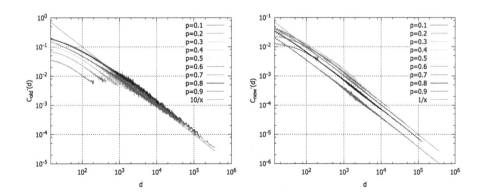

Fig. 4. Comparison of "new" and "old" parts of the average local clustering coefficient.

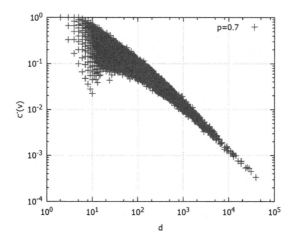

Fig. 5. The distribution of individual local clustering coefficients.

Finally, Fig. 5 shows the distribution of the individual local clustering coefficients for one graph generated with $p = 0.7$. Theorem 3 states that a.a.s. there exist a vertex v of degree d with $c^-(v, n) \gg 1/d$. Also, according to this theorem, the situation is much worse for smaller values of d. Indeed, one can see on Fig. 5 that for small d the scatter of points is much larger. On the other hand, in Theorem 4 we present bounds for $c^-(v, n)$ for almost all vertices, provided that d is large enough. One can see it on the figure too and, similarly to previously discussed figures, we observe the expected behaviour even for relatively small n despite the bound $\log^C n$ that is bigger than n in our case.

Acknowledgements. This study was funded by RFBR according to the research project 18-31-00207, Russian President grant MK-527.2017.1, and NSERC.

References

1. Aiello, W., Bonato, A., Cooper, C., Janssen, J., Prałat, P.: A spatial web graph model with local influence regions. Internet Math. **5**, 175–196 (2009)
2. Barabási, A.L., Albert, R.: Emergence of scaling in random networks. science **286**(5439), 509–512 (1999)
3. Bezanson, J., Edelman, A., Karpinski, S., Shah, V.: Julia: a fresh approach to numerical computing. SIAM Rev. **69**, 65–98 (2017)
4. Boccaletti, S., Latora, V., Moreno, Y., Chavez, M., Hwang, D.U.: Complex networks: structure and dynamics. Phys. Rep. **424**(4), 175–308 (2006)
5. Bollobás, B., Riordan, O.M.: Mathematical results on scale-free random graphs. In: Bornholdt, S., Schuster, H.G. (eds.) Handbook of Graphs and Networks: From the Genome to the Internet, pp. 1–34. Wiley, Hoboken (2003)
6. Bromberger, S., other contributors: Juliagraphs/lightgraphs.jl: Lightgraphs v0.10.5, September 2017

7. Cooper, C., Frieze, A., Prałat, P.: Some typical properties of the spatial preferred attachment model. Internet Math. **10**, 27–47 (2014)
8. Costa, L.F., Rodrigues, F.A., Travieso, G., Villas Boas, P.R.: Characterization of complex networks: a survey of measurements. Adv. Phys. **56**(1), 167–242 (2007)
9. Csányi, G., Szendrői, B.: Structure of a large social network. Phys. Rev. E **69**(3), 036131 (2004)
10. Dorogovtsev, S.N., Goltsev, A.V., Mendes, J.F.F.: Pseudofractal scale-free web. Phys. Rev. E **65**(6), 066122 (2002)
11. Faloutsos, M., Faloutsos, P., Faloutsos, C.: On power-law relationships of the internet topology. ACM SIGCOMM Comput. Commun. Rev. **29**, 251–262 (1999)
12. Iskhakov, L., Mironov, M., Ostroumova Prokhorenkova, L., Pralat, P.: Local clustering coefficient of spatial preferential attachment model. arXiv preprint arXiv:1711.06846 (2017)
13. Jacob, E., Mörters, P.: A spatial preferential attachment model with local clustering. In: Bonato, A., Mitzenmacher, M., Prałat, P. (eds.) WAW 2013. LNCS, vol. 8305, pp. 14–25. Springer, Cham (2013). https://doi.org/10.1007/978-3-319-03536-9_2
14. Jacob, E., Mörters, P., et al.: Spatial preferential attachment networks: power laws and clustering coefficients. Ann. Appl. Probab. **25**(2), 632–662 (2015)
15. Janssen, J., Hurshman, M., Kalyaniwalla, N.: Model selection for social networks using graphlets. Internet Math. **8**(4), 338–363 (2013)
16. Janssen, J., Prałat, P., Wilson, R.: Geometric graph properties of the spatial preferred attachment model. Adv. Appl. Math. **50**, 243–267 (2013)
17. Janssen, J., Prałat, P., Wilson, R.: Non-uniform distribution of nodes in the spatial preferential attachment model. Internet Math. **12**(1–2), 121–144 (2016)
18. Kamiński, B., Olczak, T., Prałat, P.: Endogenous differentiation of consumer preferences under quality uncertainty in a SPA network. In: Bonato, A., Chung Graham, F., Prałat, P. (eds.) WAW 2017. LNCS, vol. 10519, pp. 42–59. Springer, Cham (2017). https://doi.org/10.1007/978-3-319-67810-8_4
19. Krot, A., Ostroumova Prokhorenkova, L.: Local clustering coefficient in generalized preferential attachment models. In: Gleich, D.F., Komjáthy, J., Litvak, N. (eds.) WAW 2015. LNCS, vol. 9479, pp. 15–28. Springer, Cham (2015). https://doi.org/10.1007/978-3-319-26784-5_2
20. Leskovec, J.: Dynamics of large networks (proquest), Ann Arbor (2008)
21. Newman, M.E.: Properties of highly clustered networks. Phys. Rev. E **68**(2), 026121 (2003)
22. Newman, M.E.: The structure and function of complex networks. SIAM Rev. **45**(2), 167–256 (2003)
23. Newman, M.E.: Power laws, pareto distributions and Zipf's law. Contemp. Phys. **46**(5), 323–351 (2005)
24. Ostroumova Prokhorenkova, L.: Global clustering coefficient in scale-free weighted and unweighted networks. Internet Math. **12**(1–2), 54–67 (2016)
25. Ostroumova Prokhorenkova, L., Prałat, P., Raigorodskii, A.: Modularity of complex networks models. Preprint (2017)
26. Ravasz, E., Barabási, A.L.: Hierarchical organization in complex networks. Phys. Rev. E **67**(2), 026112 (2003)
27. Serrano, M.Á., Boguna, M.: Clustering in complex networks. I. General formalism. Phys. Rev. E **74**(5), 056114 (2006)
28. Vázquez, A., Pastor-Satorras, R., Vespignani, A.: Large-scale topological and dynamical properties of the internet. Phys. Rev. E **65**(6), 066130 (2002)

Parameter Estimators of Sparse Random Intersection Graphs with Thinned Communities

Joona Karjalainen[1(✉)], Johan S. H. van Leeuwaarden[2], and Lasse Leskelä[1]

[1] Aalto University, Espoo, Finland
`joona.karjalainen@aalto.fi`
[2] Eindhoven University of Technology, Eindhoven, The Netherlands

Abstract. This paper studies a statistical network model generated by a large number of randomly sized overlapping communities, where any pair of nodes sharing a community is linked with probability q via the community. In the special case with $q = 1$ the model reduces to a random intersection graph which is known to generate high levels of transitivity also in the sparse context. The parameter q adds a degree of freedom and leads to a parsimonious and analytically tractable network model with tunable density, transitivity, and degree fluctuations. We prove that the parameters of this model can be consistently estimated in the large and sparse limiting regime using moment estimators based on partially observed densities of links, 2-stars, and triangles.

1 Introduction

Networks often display transitivity or clustering, the tendency for nodes to be connected if they share a mutual neighbor. Random graphs can statistically model networks with clustering after adding a community structure of small relatively dense subgraphs. Triangles, or other short cycles, then occur predominantly within and not between the communities, and clustering becomes tunable through adapting the community structure.

There are various ways to install community structure, for instance by locally adding small dense graphs [1–4]. This creates nonoverlapping communities. Another way is to introduce overlapping communities through a random intersection graph (RIG) which can be defined as the 2-section of a random inhomogeneous hypergraph where hyperedges correspond to overlapping communities [5]. RIGs have attractive analytical features, for example admitting tunable transitivity (clustering coefficient) and power-law degree distributions [6–8]. However, by construction the RIG community structure is rigid, in the sense that every community corresponds to a clique. In this paper we relax this property and consider an extension of the RIG, a thinned RIG where nodes within the same community are linked with some probability $q \in [0, 1]$ via that community, independently across all node pairs.

A. Bonato et al. (Eds.): WAW 2018, LNCS 10836, pp. 44–58, 2018.
https://doi.org/10.1007/978-3-319-92871-5_4

The RIG and thinned RIG are known to generate high levels of transitivity, even in sparse regimes where nodes have finite mean degrees in the large-network limit [6,9]. In [9] it is shown that the community density q can be exploited to tune both triangle and 4-cycle densities. In this paper we also exploit the additional freedom offered by q, but for controlling the density of 2-stars instead of 4-cycles. We derive scaling relations between the model parameters to create large, sparse, clustered networks, in which the number of links grows linearly in the number of nodes n, and the numbers of 2-stars and triangles grow quadratically in n. We investigate a special instance of the sparse model parameterized by a triplet (λ, μ, q) where λ corresponds to the mean degree and μ to the mean number of community memberships of a node. By analyzing limiting expressions for the link, 2-star and triangle densities, we derive moment estimators for λ, μ, and q based on observed frequencies of 2-stars and triangles. Taken together, the densities of links, 2-stars and triangles prove sufficient to produce tunable sparsity (mean degree), degree fluctuations and transitivity.

This work is part of an emerging area in network science that connects high-order local network structure such as subgraphs with statistical estimation procedures. The triangle is the most studied subgraph, because it not only describes transitivity, but also signals hierarchy and community structure [10]. Other subgraphs, however, such as 2-stars, bifans, cycles, and cliques are also relevant for understanding network organization [11,12]. In this paper we exploit a direct connection between the model parameters and the frequencies of links, 2-stars and triangles. A key technical challenge is to characterize the mean and variance of the subgraph frequencies, where the latter requires frequencies of all subgraphs that can be constructed by merging two copies of the subgraph at hand [13–16]. A byproduct of our analysis yields a rigorous proof of the graph-ergodic theorem (analogous to [17, Theorem 3.2]) stating that the observed transitivity (a large graph average) of a large graph sample is with high probability close to the model transitivity (a probabilistic average).

Notation. For a probability distribution π on the nonnegative integers, we denote the moments by $\pi_r = \sum_x x^r \pi(x)$ and the factorial moments by $(\pi)_r = \sum_x (x)_r \pi(x)$, where $(x)_r = x(x-1)\cdots(x-r+1)$. For sequences a_n and b_n, we denote $a \lesssim b$ when $a_n \leq c b_n$ for some $c > 0$ and all n. $a \asymp b$ means "$a \lesssim b$ and $b \lesssim a$". For $a_n = (1 + o(1))b_n$ we use the notation $a \sim b$, and for $a_n/b_n \to 0$ we use $a \ll b$. $X = o_{\mathbb{P}}(1)$ is read as "X converges to zero in probability".

2   Model Description

We will study a statistical network model with n nodes (individuals, users, vertices) and m overlapping communities (attributes, blocks, groups, layers). The model is parameterized by (n, m, π, q), where π is a probability distribution on $\{0, \ldots, n\}$ such that $\pi(x)$ corresponds to the proportion of communities of size x, and $q \in [0, 1]$ is the probability that two nodes are linked via a particular community.

A realization of the model corresponds to a collection of random subsets V_k of $\{1, \ldots, n\}$ indexed by $k = 1, \ldots, m$ representing the communities, and a collection of symmetric binary matrices $(C_{ij,k})_{ij}$, with $i, j = 1, \ldots, n$, and $k = 1, \ldots, m$. These objects are used to define an undirected random graph G on node set $\{1, \ldots, n\}$ with adjacency matrix

$$G_{ij} = \max_{k=1,\ldots,m} \{B_{i,k} B_{j,k} C_{ij,k}\}, \quad i \neq j, \tag{1}$$

where $B_{i,k} = 1_{V_k}(i)$ indicates whether node i belongs to community k, and $C_{ij,k} = 1$ means that i and j are linked via community k, given that both i and j are members of community k. We assume that $V_1, \ldots, V_m$ are independent random sets with a common probability density $\mathbb{P}(V_i = A) = \pi(|A|)\binom{n}{|A|}^{-1}$, and that $C_{ij,k}$ are independent $\{0, 1\}$-valued random integers with mean q. Moreover, the arrays (V_k) and $(C_{ij,k})$ are assumed independent.

The special case where $q = 1$ corresponds to the so-called passive random intersection graph model [7,18]. The special case where π is a Dirac measure has been recently studied in [9]. The binomial community size distribution $\pi(x) = \binom{n}{x}(1-p)^{n-x}p^x$ gives another important special case of the model (referred to as Bernoulli model), which allows to smoothly interpolate between a standard Erdős–Rényi random graph (setting $p = 1$) and a binomial random intersection graphs [19] (with $q = 1$).

3   Analysis of Local Model Characteristics

3.1   Sparse Parameter Regime

In this section we analyze how the model behaves when the number of nodes n is large. We view a large network as a sequence of models with parameter quadruples $(n, m, \pi, q) = (n_\nu, m_\nu, \pi_\nu, q_\nu)$ indexed by a scale parameter $\nu = 1, 2, \ldots$ such that $n_\nu \to \infty$ as $\nu \to \infty$. For simplicity we omit the scale parameter from the notation.

Let $p_r = (\pi)_r / (n)_r$ denote the probability that a particular community contains a given set of r nodes. Then $m p_r$ equals the mean number of communities common to a particular set of r nodes, and $\binom{n}{r}p_r = (\pi)_r / r!$ equals the expected number of r-sets of nodes contained in a single community. Because $m p_2 q$ equals the number of communities through which a given node pair is linked, it is natural to assume that $m p_2 q \ll 1$ when modeling a large and sparse network. The following result confirms this.

Proposition 1. *The probability that any particular pair of distinct nodes is linked equals* $\mathbb{P}(\text{link}) = 1 - (1 - q p_2)^m$. *Furthermore,* $\mathbb{P}(\text{link}) \ll 1$ *if and only if* $m p_2 q \ll 1$, *in which case*

$$\mathbb{P}(\text{link}) = (1 + O(m p_2 q)) \, m p_2 q. \tag{2}$$

3.2   Subgraph Densities

For an arbitrary graph R, the R-*covering density* of the model is defined as the expected proportion of subgraphs[1] of G that are isomoprhic to R. By symmetry, this quantity equals the probability that G contains R as a subgraph, when we assume that $V(R) \subset \{1, \ldots, n\}$. Note that the K_2-covering density of the model is just the link density analyzed in Proposition 1. The following result describes the covering densities of connected three-node graphs.

Proposition 2. *The probabilities that the model in the sparse regime $mp_2q \ll 1$ contains as subgraph the 2-star and triangle are approximately*

$$\mathbb{P}(2\text{-}star) \;=\; (1 + O(mp_2q))\, q^2 \Big(mp_3 + (m)_2 p_2^2 \Big), \tag{3}$$

$$\mathbb{P}(\text{triangle}) \;=\; (1 + O(mqp_2))\, q^3 \Big(mp_3 + 3(m)_2 p_2 p_3 + (m)_3 p_2^3 \Big). \tag{4}$$

3.3   Model Transitivity

The transitivity (or global clustering coefficient) of a graph usually refers to the proportion of triangles among unordered node triplets which induce a connected graph. The *model transitivity* of a random graph is usually defined by replacing the numerator and the denominator in the latter expression by their expected values [17]. In our case, by symmetry, the model transitivity equals $\tau = \mathbb{P}(\text{triangle})/\mathbb{P}(2\text{-star})$, and is characterized by the following result in the sparse parameter regime.

Proposition 3. *The model transitivity in the sparse regime $mp_2q \ll 1$ satisfies*

$$\tau \;=\; \frac{p_3 q}{p_3 + (m-1)p_2^2} + o(1).$$

Remark 1. In the special case with $q = 1$ the above result coincides with [20, Corollary 1] and [7, Theorem 3.2].

3.4   Degree Mean and Variance

Proposition 4. *The degree D of any particular node of the model in the sparse regime $mp_2q \ll 1$ satisfies*

$$\mathbb{E}(D) \;\sim\; mnp_2q, \qquad \mathrm{Var}(D) \;\sim\; mnp_2q \left(1 + nq \left(\frac{p_3}{p_2} - p_2 \right) \right).$$

---

[1] By subgraph we mean any subgraph, not just the induced ones.

4 Parameter Estimation of Sparse Models

Our goal is to fit the model parameters to a sparse and large graph sample of known size n in a consistent way. For this we impose assumptions on the parameter sequence $(n_\nu, m_\nu, \pi_\nu, q_\nu)$, called the balanced sparse regime.

Assumption 1 (Balanced sparse regime). *The ratio m/n, the factorial moments $(\pi)_1$, $(\pi)_2$, $(\pi)_3$, and the parameter q converge to nonzero finite constants as the scale parameter tends to infinity.*

Propositions 3 and 4 imply that in the balanced sparse regime, the mean degree λ, the degree variance σ^2, and the model transitivity τ converge to nonzero finite constants which are related to the model characteristics via the formulas

$$\lambda \sim (m/n)(\pi)_2 q, \qquad \sigma^2 \sim \lambda\left(1 + q\frac{(\pi)_3}{(\pi)_2}\right), \qquad \tau \sim \frac{(\pi)_3 q}{(\pi)_3 + (m/n)(\pi)_2^2}.$$

These are the three model characteristics we wish to fit to real data. Single-parameter distributions π are of special interest, as the parameter then determines both $(\pi)_2$ and $(\pi)_3$, reducing the number of unknowns by one.

4.1 Empirical Subgraph Counts

Consider the model $G = (n, m, q, \pi)$ and assume that we have observed a subgraph $G^{(n_0)}$ induced by n_0 nodes. We wish to estimate one or more model parameters using the empirical subgraph counts in $G^{(n_0)}$ and the asymptotic relations developed in Sect. 3. Computationally efficient estimators are obtained by choosing a suitably low n_0.

Denote by $N_{K_2}(G^{(n_0)})$ the number of links, by $N_{S_2}(G^{(n_0)})$ the number of (induced or noninduced) subgraphs which are isomorphic to the 2-star, and by $N_{K_3}(G^{(n_0)})$ the number of triangles in the observed graph $G^{(n_0)}$. These are asymptotically close to the expected subgraph counts by the following theorem.

Theorem 2. *Consider the model in the balanced sparse regime* (Assumption 1). *If $(\pi)_4 \lesssim 1$ and $n_0 \gg n^{1/2}$, then the number of links in the observed graph $G^{(n_0)}$ satisfies*

$$N_{K_2}(G^{(n_0)}) = (1 + o_{\mathbb{P}}(1))\mathbb{E}N_{K_2}(G^{(n_0)}). \tag{5}$$

If also $(\pi)_6 \lesssim 1$, and $n_0 \gg n^{2/3}$, then

$$N_{S_2}(G^{(n_0)}) = (1 + o_{\mathbb{P}}(1))\,\mathbb{E}N_{S_2}(G^{(n_0)}), \tag{6}$$

$$N_{K^3}(G^{(n_0)}) = (1 + o_{\mathbb{P}}(1))\,\mathbb{E}N_{K_3}(G^{(n_0)}). \tag{7}$$

4.2   Parameter Estimation in the Bernoulli Model

The binomial community size distribution $\pi(x) = \binom{n}{x}(1-p)^{n-x}p^x$ with $p \in (0,1)$ gives $(\pi)_r = n!/(n-r)!p^r$ for all integers $r \geq 1$. We parameterize the model with three positive constants (λ, μ, q) (with q not depending on scale) and choose

$$m = \left\lfloor \frac{\mu^2 q}{\lambda} n \right\rfloor, \quad \text{and} \quad p = \frac{\lambda}{\mu q} n^{-1}, \tag{8}$$

where μ can be interpreted as the mean number of communities of a node. The following (asymptotic) relations follow from the results in Sect. 3:

$$\lambda \sim nqmp^2, \quad \sigma^2 \sim nqmp^2(1+nqp), \quad \tau \sim \frac{q}{1+mp},$$

from which one may solve

$$\mu = \frac{\lambda^2}{\sigma^2 - \lambda} \quad \text{and} \quad q = \tau(1 + \frac{\lambda^2}{\sigma^2 - \lambda}).$$

After substituting the asymptotic densities from Sect. 3 and estimating them using empirical counts we obtain (after some algebra) the estimators

$$\hat{\lambda} = (n-1)\binom{n_0}{2}^{-1} N_{K_2}(G^{(n_0)}),$$

$$\hat{\mu} = \frac{2N_{K_2}(G^{(n_0)})^2}{n_0 N_{S_2}(G^{(n_0)}) - 2N_{K_2}(G^{(n_0)})^2}, \quad \hat{q} = \frac{3n_0 N_{K_3}(G^{(n_0)})}{n_0 N_{S_2}(G^{(n_0)}) - 2N_{K_2}(G^{(n_0)})^2}.$$

To summarize, we estimate the parameters μ and q by counting the numbers of links, 2-stars, and triangles from an induced subgraph of n_0 nodes. Alternatively, this can be seen as a way of fitting the transitivity and the mean and variance of the degrees. The theoretical justification is given by the following theorem.

Theorem 3. $\hat{\lambda}$, $\hat{\mu}$, and $\hat{q}$ converge in probability to the true values λ, μ, and q, under the Bernoulli model defined by (8) given $n_0 \gg n^{2/3}$.

Proof. The assumptions of Theorem 2 and Propositions 1 and 4 are satisfied by (8), which establishes the claim for $\hat{\lambda}$. Dividing and multiplying both $\hat{\mu}$ and $\hat{q}$ by n_0^2 yields rational expressions where the numerators and denominators converge in probability to nonzero constants by Theorem 2 and Propositions 1 and 2. The claim now follows from the continuous mapping theorem.

5   Numerical Experiments

5.1   Attainable Regions in the Bernoulli Model

The relations $\sigma^2 \geq \lambda$, $\tau \in (0,1)$ and $\tau \leq (1 + \lambda^2/(\sigma^2 - \lambda))^{-1}$ restrict the attainable combinations $(\lambda, \tau, \sigma^2)$; see Fig. 1. To obtain a model with a large asymptotic transitivity coefficient, one may choose a low mean degree and a large degree variance. The flexibility gained by allowing $q \leq 1$ is also illustrated in Fig. 2. The discreteness of the attainable points $(\mathbb{P}(\text{link}), \mathbb{P}(\text{triangle}))$ is obvious with $q = 1$, whereas the points with $q \leq 1$ fill a large part above the curve $\mathbb{P}(\text{triangle}) = \mathbb{P}(\text{link})^3$.

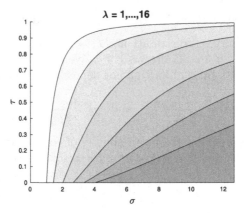

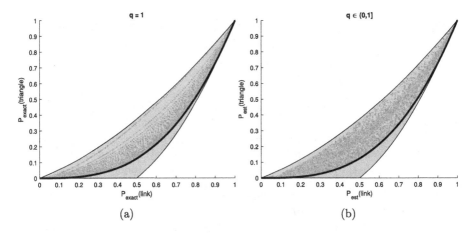

Fig. 1. Attainable combinations of (τ, σ) for $\lambda \in \{1, 2, 4, 7, 11, 16\}$. Combinations with $q = 1$ lie on the curves. The points under the curves are obtained by setting $q \leq 1$.

(a)   (b)

Fig. 2. Attainable combinations of link and triangle probabilities in Bernoulli models with different values of λ and τ ($\lambda \leq 500$, $\tau \geq 0.0002$), and (a) $q = 1$ (exact probabilities) and (b) $q \leq 1$ (averages of 1000 Monte Carlo samples). The solid curves represent theoretical bounds, and the thick black curve the Erdős–Rényi graph.

5.2 Real Data

Ten data sets of different sizes were analyzed using the Bernoulli model. The whole data sets were used for estimation, i.e., $n_0 = n$. The obtained estimates are listed in Table 1. Because we essentially fit τ and λ, these values are listed in Table 1 only for illustration purposes. In the largest data sets the estimates of q are very small, which might suggest that the structure of the model is not strongly supported by the data. For one of the data sets, Dolphin, the estimate of q is outside the allowed range $(0, 1)$. This may be related to the denseness of the network. On the other

Table 1. Parameter estimates of the Bernoulli model for collaboration networks in astrophysics and high energy physics, a social network of bottleneck dolphins, an e-mail network from a research institution, a geographically local Facebook network, a Flickr image network, a social network of Flixster users, a Twitter network of users who mention each other in their tweets, a US aiport network, and a Wikipedia communications network. Data sets from [a][21], [b][22], and [c][23].

Data set	n	$\hat{\lambda}$	$\hat{\tau}$	$\hat{q}$	$\hat{m}$	$\hat{\sigma}$	$\hat{m}_{q=1}$	$\hat{\sigma}_{q=1}$
ca-AstroPh[a]	18772	21.1	0.32	0.47	100	30.6	4092	15.1
ca-HepPh[a]	12008	19.7	0.66	0.78	15	46.6	162	28.2
Dolphin[b]	62	5.1	0.31	$-(2.36)$	1255	3.0	61	4.1
email-Eu-core[a]	1005	32.0	0.27	0.47	8	37.0	236	20.1
Facebook[b]	63731	25.6	0.15	0.21	90	40.0	82756	11.8
Flickr[a]	105938	43.7	0.40	0.46	23	115.6	5377	36.4
Flixster[b]	2523386	6.3	0.01	0.014	4	36.6	$2.1*10^9$	2.6
Twitter	2919613	8.8	0.006	<0.001	77	20.9	$9.3*10^9$	3.0
USAir97[c]	332	12.8	0.40	0.56	2	20.1	60.1	11.0
wiki-talk[a]	2394385	3.9	0.002	0.002	<1	102.5	$1.27*10^{11}$	2.0

hand, simulation results in [17] suggest that the size $n = 62$ may not be sufficient for estimators based on asymptotic moment equations.

The rightmost two columns in Table 1 display reference values of m and σ estimated for the RIG model ($q = 1$) using the estimators introduced in [17]. These estimators give very large values for m and grossly underestimate σ in the largest data sets. These observations speak for the significantly improved model fit when using the thinned RIG model instead of the classical RIG model.

6 Technical Proofs

6.1 Analysis of Link Density

Proof (Proof of Proposition 1). The probability of the event $\mathcal{E}_k$ that nodes 1 and 2 are linked via community k can be written as

$$\mathbb{P}(\mathcal{E}_k) = \mathbb{P}(V_k \supset \{i,j\}, C_{12,k} = 1) = p_2 q.$$

Because the events $\mathcal{E}_1, \ldots, \mathcal{E}_m$ are independent, it follows that

$$\mathbb{P}(\text{link}) = \mathbb{P}\left(\bigcup_k \mathcal{E}_k\right) = 1 - \prod_k \mathbb{P}(\mathcal{E}_k^c) = 1 - (1 - p_2 q)^m.$$

The inequality $1 - x \leq e^{-x}$ and the union bound $\mathbb{P}(\cup_k \mathcal{E}_k) \leq \sum_k \mathbb{P}(\mathcal{E}_k)$ imply that $1 - e^{-mp_2 q} \leq \mathbb{P}(\text{link}) \leq mp_2 q$, from which we see that $\mathbb{P}(\text{link}) \ll 1$ if and only if $mp_2 q \ll 1$. The approximation formula (2) follows from the Bonferroni's bounds

$$mp_2 q - \binom{m}{2}(p_2 q)^2 \leq \mathbb{P}(\text{link}) \leq mp_2 q.$$

6.2   Analysis of 2-Star Covering Density

Proof (Proof of Proposition 2: Eq. (3)). Consider a 2-star with node set $\{1, 2, 3\}$ and link set $\{\{1, 2\}, \{1, 3\}\}$. Denote by $\mathcal{B}_{A,k} = \{V_k \supset A\}$ the event that community k covers a node set A, and by $\mathcal{C}_{ij,k}$ the event that $C_{ij,k} = 1$. Then $\mathcal{E}_{ij,k} = \mathcal{B}_{ij,k} \cap \mathcal{C}_{ij,k}$ is the event that node pair ij is linked by community k. Then the probability that G contains the 2-star as a subgraph is given by

$$\mathbb{P}(\text{2-star}) = \mathbb{P}\Big(\bigcup_{k \in [m]^2} \mathcal{F}_k \Big),$$

where $\mathcal{F}_k = \mathcal{E}_{12,k_1} \cap \mathcal{E}_{13,k_2}$ for an ordered community pair $k = (k_1, k_2)$. Observe that $\mathbb{P}(\mathcal{F}_k) = q^2 p_3$ for $k_1 = k_2$ and $\mathbb{P}(\mathcal{F}_k) = q^2 p_2^2$ otherwise. Therefore,

$$\mathbb{P}(\text{2-star}) \leq \sum_{k \in [m]^2} \mathbb{P}(\mathcal{F}_k) = mq^2 p_3 + (m)_2 q^2 p_2^2.$$

To prove the claim using Bonferroni's bounds, it suffices to show that

$$\sum_{(k,\ell)} \mathbb{P}(\mathcal{F}_k, \mathcal{F}_\ell) \ll q^2 \Big(mp_3 + (m)_2 p_2^2 \Big), \tag{9}$$

where the sum on the left is over all (k, ℓ)-pairs with $k, \ell \in [m]^2$ and $k \neq \ell$.

We will now compute the sum on the left side of (9). Note that

$$\mathbb{P}(\mathcal{F}_k, \mathcal{F}_\ell) = q^{|\{k_1, \ell_1\}| + |\{k_2, \ell_2\}|} \mathbb{P}(\mathcal{B}_{12,k_1}, \mathcal{B}_{13,k_2}, \mathcal{B}_{12,\ell_1}, \mathcal{B}_{13,\ell_2}).$$

Therefore, for example, for a (k, ℓ)-pair of the form $(k_1, k_2, \ell_1, \ell_2) = (a, a, b, c)$ with distinct a, b, c we have

$$\mathbb{P}(\mathcal{F}_k, \mathcal{F}_\ell) = q^4 \mathbb{P}(\mathcal{B}_{123,a}, \mathcal{B}_{12,b}, \mathcal{B}_{13,c}) = q^4 p_2^2 p_3.$$

The table below displays the values of $\mathbb{P}(\mathcal{F}_k, \mathcal{F}_\ell)$ for all combinations of $k \neq \ell$, and the cardinalities of such combinations.

$(k_1, k_2, \ell_1, \ell_2)$	Cardinality	$\mathbb{P}(\mathcal{F}_k, \mathcal{F}_\ell)$
(a, b, c, d)	$(m)_4$	$q^4 p_2^4$
(a, b, a, c) or (a, b, c, b)	$2(m)_3$	$q^3 p_2^3$
(a, a, b, c) or (a, b, c, c) or (a, b, c, a) or (a, b, b, c)	$4(m)_3$	$q^4 p_2^2 p_3$
(a, a, b, b) or (a, b, b, a)	$2(m)_2$	$q^4 p_3^2$
(a, a, a, b) or (a, a, b, a) or (a, b, a, a) or (b, a, a, a)	$4(m)_2$	$q^3 p_2 p_3$

As a consequence,

$$\sum_{(k,\ell)} \mathbb{P}(\mathcal{F}_k, \mathcal{F}_\ell) = (m)_4 q^4 p_2^4 + 2(m)_3 q^3 p_2^3 + 4(m)_3 q^4 p_2^2 p_3 + 2(m)_2 q^4 p_3^2 + 4(m)_2 q^3 p_2 p_3$$

By noting that $p_3 \leq p_2$ and $mp_2 q \ll 1$, we see that the first three terms on the right are bounded from above by $4(mp_2 q)q^2(m)_2 p_2^2$, and the last two terms on the right are bounded from above by $4(mp_2 q)q^2 mp_3$. Hence the above sum is at most $12(mp_2 q)q^2(mp_3 + (m)_2 p_2^2)$, claim (9) is valid, and the claim follows.

6.3   Analysis of Triangle Covering Density

Proof (Proof sketch of Proposition 2: Eq. (4)). Consider a triangle with node set $\{1, 2, 3\}$. Denote by $\mathcal{E}_{e,k} = \{V_k \supset e, C_{e,k} = 1\}$ the event that node pair e is linked via community k. Then $\mathbb{P}(\text{triangle}) = \mathbb{P}(\cup_{k \in [m]^3} \mathcal{F}_k)$, where $\mathcal{F}_k = \mathcal{E}_{12,k_1} \cap \mathcal{E}_{13,k_2} \cap \mathcal{E}_{23,k_3}$ is the event that the node pairs of the triangle are linked via communities of the triplet $k = (k_1, k_2, k_3)$. Because

$$
\mathbb{P}(\mathcal{F}_k) = q^3 \mathbb{P}(V_{k_1} \supset 12, V_{k_2} \supset 13, V_{k_3} \supset 23) = \begin{cases} q^3 p_3, & |\{k_1, k_2, k_3\}| = 1, \\ q^3 p_2 p_3, & |\{k_1, k_2, k_3\}| = 2, \\ q^3 p_2^3, & |\{k_1, k_2, k_3\}| = 3, \end{cases}
$$

the union bound implies that

$$
\mathbb{P}(\text{triangle}) \leq \sum_k \mathbb{P}(\mathcal{F}_k) \leq q^3 \left(m p_3 + 3(m)_2 p_2 p_3 + (m)_3 p_2^3 \right).
$$

By similar techniques as in the proof of (3), one can show that

$$
\sum_{(k,\ell):k \neq \ell} \mathbb{P}(\mathcal{F}_k, \mathcal{F}_\ell) \lesssim (mqp_2) \sum_k \mathbb{P}(\mathcal{F}_k) \ll \sum_k \mathbb{P}(\mathcal{F}_k),
$$

and the claim follows by Bonferroni's bounds. (The details of the lengthy computations are omitted.)

6.4   Analysis of Model Transitivity

Proof (Proof of Proposition 3). By applying Propositions 2 we find that

$$
\tau = (1 + o(1))q \frac{mp_3 + 3(m)_2 p_2 p_3 + (m)_3 p_2^3}{mp_3 + (m)_2 p_2^2} = (1 + o(1))q \left(\frac{mp_3}{mp_3 + (m)_2 p_2^2} + R \right),
$$

where

$$
R = \frac{3(m)_2 p_2 p_3 + (m)_3 p_2^3}{mp_3 + (m)_2 p_2^2} \leq mp_2 \frac{3mp_3 + (m)_2 p_2^2}{mp_3 + (m)_2 p_2^2} \leq 3mp_2.
$$

The assumption $mqp_2 \ll 1$ now implies that $qR = o(1)$. Hence we conclude

$$
\tau = (1 + o(1)) \left(\frac{qmp_3}{mp_3 + (m)_2 p_2^2} + o(1) \right) = \frac{qp_3}{p_3 + (m-1)p_2^2} + o(1).
$$

6.5   Analysis of Degree Moments

Proof (Proof of Proposition 4). By expressing the degree of node i using the adjacency matrix as $D = \sum_{j \neq i} G_{i,j}$ and taking expectations, we find that

$$
\mathbb{E}(D) = (n-1)\mathbb{P}(\text{link}),
$$
$$
\mathbb{E}(D^2) = (n-1)\mathbb{P}(\text{link}) + (n-1)(n-2)\mathbb{P}(\text{2-star}).
$$

By Propositions 1 and 2 we find that

$$\mathbb{P}(\text{link}) = (1 + O(mp_2q))mp_2q,$$

$$\mathbb{P}(\text{2-star}) - \mathbb{P}(\text{link})^2 = (1 + O(mp_2q))q^2\left(mp_3 + (m)_2 p_2^2 - m^2 p_2^2\right).$$

Hence $\mathbb{E}(D) \sim mnp_2q$, and by the formula $\text{Var}(D) = \mathbb{E}(D^2) - (\mathbb{E}D)^2$,

$$\text{Var}(D) = (1 + O(n^{-1}))\left(n\mathbb{P}(\text{link}) + n^2\left(\mathbb{P}(\text{2-star}) - \mathbb{P}(\text{link})^2\right)\right)$$

$$= (1 + O(n^{-1}))(1 + O(mp_2q))\left(mnqp_2 + mn^2q^2(p_3 - p_2^2)\right).$$

6.6   Analysis of Observed Link Density

Proof (Proof of Theorem 2: Eq. (5)). Let us denote by $\hat{N} = N_{K_2}(G^{(n_0)})$ the number of links in the observed graph $G^{(n_0)}$. The assumptions $(\pi)_2 \gtrsim 1$ and $(\pi)_4 \lesssim 1$ imply that $p_2 \asymp n^{-2}$ and $p_r \lesssim n^{-r}$ for $r = 3, 4$. Because $m \asymp n$, and $q \gtrsim 1$, with the help of Proposition 1, we see that

$$\mathbb{P}(\text{link}) = (1 + o(1))mp_2q \asymp n^{-1},$$

and

$$\mathbb{E}\hat{N} = \binom{n_0}{2}\mathbb{P}(\text{link}) \asymp n_0^2 n^{-1} \gg 1.$$

Denote by $\mathbb{P}(\text{link}^2)$ the probability that G contains any particular pair of disjoint node pairs (e.g., pairs {1,2} and {3,4}). Note that

$$\text{Var}(\hat{N}) = \sum_e \sum_{e'} \mathbb{P}(e \in E(G^{(n_0)}), e' \in E(G^{(n_0)})) - \binom{n_0}{2}^2 \mathbb{P}(\text{link})^2$$

$$= \binom{n_0}{2}\mathbb{P}(\text{link}) + (n_0)_3 \mathbb{P}(\text{2-star}) + \binom{n_0}{2}\binom{n_0 - 2}{2}\mathbb{P}(\text{link}^2) - \binom{n_0}{2}^2 \mathbb{P}(\text{link})^2$$

$$\leq n_0^2 \mathbb{P}(\text{link}) + n_0^3 \mathbb{P}(\text{2-star}) + \binom{n_0}{2}^2\left(\mathbb{P}(\text{link}^2) - \mathbb{P}(\text{link})^2\right).$$

Note that $\mathbb{P}(\text{link}) \asymp n^{-1}$ and $\mathbb{P}(\text{2-star}) \lesssim n^{-2}$. Furthermore,

$$\mathbb{P}(\text{link}^2) = \mathbb{P}(\cup_k \cup_\ell \{V_k \supset \{1,2\}, C_{12,k} = 1, V_\ell \supset \{3,4\}, C_{34,\ell} = 1\})$$

$$\leq \sum_k \sum_\ell \mathbb{P}(\{V_k \supset \{1,2\}, C_{12,k} = 1, V_\ell \supset \{3,4\}, C_{34,\ell} = 1\})$$

$$= (m)_2 p_2^2 q^2 + mp_4 q^2 = (1 + o(1))\mathbb{P}(\text{link})^2 + O(n^{-3}),$$

so that

$$\text{Var}(\hat{N}) \lesssim n_0^2 n^{-1} + n_0^3 n^{-2} + n_0^4 n^{-3} + o(1)(\mathbb{E}\hat{N})^2$$

$$\leq 3n_0^2 n^{-1} + o(1)(\mathbb{E}\hat{N})^2 \ll (\mathbb{E}\hat{N})^2.$$

6.7 Analysis of Observed 2-Star Covering Density

Proof (Proof sketch of Theorem 2: Eq. (6)). Let us denote $\hat{N} = N_{S_2}(G^{(n_0)})$. Note that

$$\hat{N} = \sum_R 1_{A_R}$$

where the sum ranges over the set of all S_2-isomorphic subgraphs of $K_{[n_0]}$, and 1_{A_R} is the indicator of the event A_R that $G^{(n_0)}$ contains R as a subgraph. The assumptions $(\pi)_2 \gtrsim 1$ and $(\pi)_6 \lesssim 1$ imply that $p_2 \asymp n^{-2}$ and $p_r \lesssim n^{-r}$ for $r = 3,\ldots,6$. Because $m \asymp n$, and $q \gtrsim 1$, with the help of Proposition 2, we see that

$$\mathbb{P}(\text{2-star}) = q^2 \left(mp_3 + (m)_2 p_2^2 \right)(1 + o(1)) \asymp n^{-2}, \tag{10}$$

and

$$\mathbb{E}\hat{N} = 3\binom{n_0}{3}\mathbb{P}(\text{2-star}) \asymp n_0^3 n^{-2} \gg 1.$$

The above relation underlines the role of assumption $n_0 \gg n^{2/3}$. This guarantees that there are lots of (dependent) samples to sum in the observed graph.

Let us next analyze the variance of $\hat{N}$. By applying the formula $\text{Var}(\hat{N}) = \mathbb{E}(\hat{N}^2) - (\mathbb{E}\hat{N})^2$ and noting that $A_R \cap A_{R'} = A_{R \cup R'}$, we see that

$$\text{Var}(\hat{N}) = \sum_R \sum_{R'} \mathbb{P}(A_R, A_{R'}) - \sum_R \sum_{R'} \mathbb{P}(A_R)\mathbb{P}(A_{R'}) = \sum_{i=0}^3 M_i,$$

where

$$M_i = \sum_R \sum_{R':|V(R)\cap V(R')|=i} \left(\mathbb{P}(A_{R\cup R'}) - \mathbb{P}(A_R)^2 \right). \tag{11}$$

For $i \geq 1$, we approximate M_i from above by omitting the $\mathbb{P}(A_R)$ term in (11). By generalizing the analytical technique used in [17] (details will be available in the extended version), it can be shown that for any graph R such that $|V(R)| \leq 6$,

$$\mathbb{P}(A_R) \lesssim n^{-\kappa(R)}, \tag{12}$$

where $\kappa(R) = \min_{\mathcal{E}}(||\mathcal{E}|| - |\mathcal{E}|)$, with the minimum taken across all partitions of $E(R)$ into nonempty sets, where $|\mathcal{E}|$ is the number of parts in the partition, and we set $||\mathcal{E}|| = \sum_{E \in \mathcal{E}} |E^\flat|$ where $E^\flat = \cup_{e \in E} e$ denotes the set of nodes covered by the node pairs of E, so that for example, $\{\{1,2\}\}^\flat = \{1,2\}$ and $\{\{1,3\},\{2,3\}\}^\flat = \{1,2,3\}$. Table 2 summarizes the values of $\kappa(R)$ for the type of graphs that can be obtained as unions of two 2-stars. By applying (12), it follows that

$$M_1 \lesssim n_0^5 \left(\mathbb{P}(\text{4-star}) + \mathbb{P}(\text{4-path}) + \mathbb{P}(\text{chair}) \right) \lesssim n_0^5 n^{-4},$$

$$M_2 \lesssim n_0^4 \left(\mathbb{P}(\text{3-star}) + \mathbb{P}(\text{3-path}) + \mathbb{P}(\text{3-pan}) + \mathbb{P}(\text{4-cycle}) \right) \lesssim n_0^4 n^{-3},$$

$$M_3 \lesssim n_0^3 \left(\mathbb{P}(\text{2-star}) + \mathbb{P}(\text{triangle}) \right) \lesssim n_0^3 n^{-2}.$$

Table 2. Values of $\kappa(R)$ (obtained using an exhaustive computer search) for graphs obtained as unions of two 2-stars.

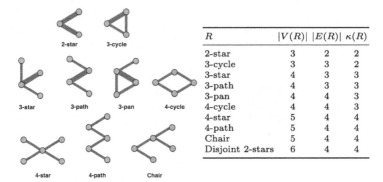

R	$|V(R)|$	$|E(R)|$	$\kappa(R)$
2-star	3	2	2
3-cycle	3	3	2
3-star	4	3	3
3-path	4	3	3
3-pan	4	4	3
4-cycle	4	4	3
4-star	5	4	4
4-path	5	4	4
Chair	5	4	4
Disjoint 2-stars	6	4	4

Because $n_0 \gg n^{2/3}$, it follows that $M_i \lesssim n_0^3 n^{-2} \ll (\mathbb{E}\hat{N})^2$ for $i = 1, 2, 3$.

The M_0-term in the variance formula (11) satisfies

$$M_0 \asymp n_0^6 \left(\mathbb{P}(\text{2-star}^2) - \mathbb{P}(\text{2-star})^2 \right)$$

where $\mathbb{P}(\text{2-star}^2)$ indicates the probability that G contains a particular union of two disjoint 2-stars as a subgraph. Here we need more careful analysis because the technique used to bound M_i for $i \geq 1$ would only yield an upper bound for M_0 of the same order as $(\mathbb{E}\hat{N})^2$. Nevertheless, a tedious but straightforward computation (details will be available in the extended version) involving all 15 partitions of the link set of a union of two disjoint 2-stars can be used to verify that

$$\mathbb{P}(\text{2-star}^2) \leq q^4 \left(m^2 p_3^2 + 2m^3 p_2^2 p_3 + m^4 p_2^4 \right) + O(n^{-5})$$
$$= (1 + o(1))\mathbb{P}(\text{2-star})^2 + O(n^{-5}).$$

By comparing this with (10), we find that $\mathbb{P}(\text{2-star}^2) - \mathbb{P}(\text{2-star})^2 \ll \mathbb{P}(\text{2-star})^2$, and

$$M_0 \asymp n_0^6 \left(\mathbb{P}(\text{2-star}^2) - \mathbb{P}(\text{2-star})^2 \right) \ll n_0^6 \mathbb{P}(\text{2-star})^2 \asymp (\mathbb{E}\hat{N})^2.$$

We may now conclude that $\text{Var}(\hat{N}) = \sum_{i=0}^3 M_i \ll (\mathbb{E}\hat{N})^2$, and hence the claim follows by Chebyshev's inequality.

6.8   Analysis of Observed Triangle Density

Proof (Proof sketch of Theorem 2: Eq. (7)). Let us denote by $\hat{N} = N_{K_3}(G^{(n_0)})$ the number of triangles in the observed graph $G^{(n_0)}$. The assumptions $(\pi)_2 \gtrsim 1$ and $(\pi)_6 \lesssim 1$ imply that $p_2 \asymp n^{-2}$ and $p_r \lesssim n^{-r}$ for $r = 3, \ldots, 6$. Because $m \asymp n$, and $q \gtrsim 1$, with the help of Proposition 2, we see that

$$\mathbb{P}(\text{triangle}) = (1 + o(1))mp_3 q^3 \asymp n^{-2},$$

and

$$\mathbb{E}\hat{N} = \binom{n_0}{3}\mathbb{P}(\text{triangle}) \asymp n_0^3 n^{-2} \gg 1.$$

To show that $\hat{N}$ is with high probability close to $\mathbb{E}\hat{N}$, by Chebyshev's inequality it suffices to verify that $\mathrm{Var}(\hat{N}) \ll (\mathbb{E}\hat{N})^2$. By applying the formula $\mathrm{Var}(\hat{N}) = \mathbb{E}\hat{N}^2 - (\mathbb{E}\hat{N})^2$ and noting that $A_R \cap A_{R'} = A_{R \cup R'}$, we see that

$$\mathrm{Var}(\hat{N}) = \sum_R \sum_{R'} \mathbb{P}(A_R, A_{R'}) - \sum_R \sum_{R'} \mathbb{P}(A_R)\mathbb{P}(A_{R'}) = \sum_{i=0}^{3} M_i,$$

where

$$M_i = \sum_R \sum_{R':|V(R)\cap V(R')|=i} \left(\mathbb{P}(A_{R\cup R'}) - \mathbb{P}(A_R)^2 \right).$$

In analogy with the proof of (6) one can show (details omitted) that $M_i \ll (\mathbb{E}\hat{N})^2$ for $i = 1, 2, 3$ by analyzing the subgraph containment probabilities of G for unions of two triangles. Again, the M_0 term requires special attention. A careful analysis of the various patterns through which the communities of the model can cover the links of two disjoint triangles (details available in the extended version) shows that

$$\mathbb{P}(\text{triangle}^2) - \mathbb{P}(\text{triangle})^2 \ll \mathbb{P}(\text{triangle})^2.$$

This implies $M_0 \ll (\mathbb{E}\hat{N})^2$ and allows to conclude that $\mathrm{Var}(\hat{N}) = \sum_{i=0}^{3} M_i \ll (\mathbb{E}\hat{N})^2$. Hence the claim follows by Chebyshev's inequality.

References

1. Ball, F., Britton, T., Sirl, D.: A network with tunable clustering, degree correlation and degree distribution, and an epidemic thereon. J. Math. Biol. **66**(4), 979–1019 (2013)
2. Coupechoux, E., Lelarge, M.: How clustering affects epidemics in random networks. Adv. Appl. Probab. **46**(4), 985–1008 (2014)
3. Stegehuis, C., van der Hofstad, R., van Leeuwaarden, J.S.H.: Epidemic spreading on complex networks with community structures. Sci. Rep. **6**, 29748 (2016)
4. van der Hofstad, R., van Leeuwaarden, J.S.H., Stegehuis, C.: Hierarchical configuration model. arXiv:1512.08397 (2015)
5. Karoński, M., Scheinerman, E.R., Singer-Cohen, K.B.: On random intersection graphs: the subgraph problem. Combin. Probab. Comput. **8**(1–2), 131–159 (1999)
6. Deijfen, M., Kets, W.: Random intersection graphs with tunable degree distribution and clustering. Probab. Eng. Inform. Sc. **23**(4), 661–674 (2009)
7. Bloznelis, M.: Degree and clustering coefficient in sparse random intersection graphs. Ann. Appl. Probab. **23**(3), 1254–1289 (2013)
8. Bloznelis, M., Leskelä, L.: Diclique clustering in a directed random graph. In: Bonato, A., Graham, F.C., Prałat, P. (eds.) WAW 2016. LNCS, vol. 10088, pp. 22–33. Springer, Cham (2016). https://doi.org/10.1007/978-3-319-49787-7_3

9. Petti, S., Vempala, S.: Random overlapping communities: Approximating motif densities of large graphs. arXiv:1709.09477 (2017)
10. Ravasz, E., Barabási, A.L.: Hierarchical organization in complex networks. Phys. Rev. E **67**, 026112 (2003)
11. Benson, A.R., Gleich, D.F., Leskovec, J.: Higher-order organization of complex networks. Science **353**(6295), 163–166 (2016)
12. Tsourakakis, C.E., Pachocki, J., Mitzenmacher, M.: Scalable motif-aware graph clustering. In: Proceedings of the 26th International Conference on World Wide Web, WWW 2017, pp. 1451–1460. International World Wide Web Conferences Steering Committee, Republic and Canton of Geneva (2017)
13. Frank, O.: Moment properties of subgraph counts in stochastic graphs. Ann. N. Y. Acad. Sci. **319**(1), 207–218 (1979)
14. Picard, F., Daudin, J.J., Koskas, M., Schbath, S., Robin, S.: Assessing the exceptionality of network motifs. J. Comput. Biol. **15**(1), 1–20 (2008)
15. Matias, C., Schbath, S., Birmelé, E., Daudin, J.J., Robin, S.: Network motifs: mean and variance for the count. Revstat **4**(1), 31–51 (2006)
16. Ostilli, M.: Fluctuation analysis in complex networks modeled by hidden-variable models: necessity of a large cutoff in hidden-variable models. Phys. Rev. E **89**, 022807 (2014)
17. Karjalainen, J., Leskelä, L.: Moment-Based Parameter Estimation in Binomial Random Intersection Graph Models. In: Bonato, A., Graham, F.C., Prałat, P. (eds.) WAW 2017. LNCS, vol. 10519, pp. 1–15. Springer, Cham (2017). https://doi.org/10.1007/978-3-319-67810-8_1
18. Godehardt, E., Jaworski, J.: Two models of random intersection graphs and their applications. Electron. Notes Discrete Math. **10**, 129–132 (2001)
19. Frieze, A., Karoński, M.: Introduction to Random Graphs. Cambridge University Press, Cambridge (2016)
20. Godehardt, E., Jaworski, J., Rybarczyk, K.: Clustering coefficients of random intersection graphs. In: Gaul, W.A., Geyer-Schulz, A., Schmidt-Thieme, L., Kunze, J. (eds.) Proceedings of the 34th Annual Conference of the Gesellschaft für Klassifikation. STUDIES CLASS, pp. 243–253. Springer, Heidelberg (2012). https://doi.org/10.1007/978-3-642-24466-7_25
21. Leskovec, J., Krevl, A.: SNAP Datasets: Stanford large network dataset collection, June 2014. http://snap.stanford.edu/data
22. Kunegis, J.: Konect: the Koblenz network collection. In: Proceedings of the 22nd International Conference on World Wide Web, pp. 1343–1350. ACM (2013)
23. Batagelj, V., Mrvar, A.: Pajek datasets (2006)

Joint Alignment from Pairwise Differences with a Noisy Oracle

Michael Mitzenmacher[1] and Charalampos E. Tsourakakis[1,2(✉)]

[1] Harvard University, Cambridge, USA
michaelm@eecs.harvard.edu
[2] Boston University, Boston, USA
ctsourak@bu.edu

Abstract. In this work we consider the problem of recovering n discrete random variables $x_i \in \{0, \ldots, k-1\}, 1 \leq i \leq n$ with the smallest possible number of queries to a noisy oracle that returns for a given query pair (x_i, x_j) a noisy measurement of their modulo k pairwise difference, i.e., $y_{ij} = x_i - x_j \pmod{k}$. This is a joint discrete alignment problem with important applications in computer vision [12,23], graph mining [20], and spectroscopy imaging [22]. Our main result is a recovery algorithm (up to some offset) that solves with high probability the *non-convex* maximum likelihood estimation problem using $O(n^{1+o(1)})$ queries.

1 Introduction

Learning a joint alignment from pairwise differences is a problem with various important applications in computer vision [12,23], graph mining [20], and spectroscopy imaging [22]. Formally, there exists a set $V = [n]$ of n discrete items, and an assignment $g : V \to [k]$ according to which each item is assigned one out of k possible values. The assignment function g is unknown, but we obtain a set of pairwise noisy difference samples $\{y_{i,j} \stackrel{\text{def}}{=} g(i) - g(j) \pmod{k}\}_{(i,j) \in \Omega}$, where $\Omega \subseteq \binom{[n]}{2}$ is a symmetric index set. To give an example, imagine a set of n images of the same object, where each $g(i)$ is the orientation/angle of the camera when taking the i-th image. Recovering g would allow to better understand the 3d structure of the object. The goal is to recover g based on these measurements, up to some global offset that is unrecoverable. However, learning a joint alignment from such differences is a non-convex problem by nature, since the input space is discrete and already non-convex to begin with [5].

Model. Suppose that there are k groups, where k is a positive constant, that we number $\{0, 1, ..., k-1\}$ and that we think of as being arranged modulo k. Let $g(u)$ refer to the group number associated with a vertex u. We are allowed to query a given pair of nodes only once. When we query an edge $e = (x, y)$, we obtain

$$\tilde{f}(e) = \begin{cases} g(x) - g(y) \bmod k, & \text{with probability } 1 - q; \\ g(x) - g(y) + 1 \bmod k, & \text{with probability } q/2; \\ g(x) - g(y) - 1 \bmod k, & \text{with probability } q/2. \end{cases} \qquad (1)$$

© Springer International Publishing AG, part of Springer Nature 2018
A. Bonato et al. (Eds.): WAW 2018, LNCS 10836, pp. 59–69, 2018.
https://doi.org/10.1007/978-3-319-92871-5_5

That is, we obtain the difference between the groups when no error occurs, and with probability q we obtain an error that adds or subtracts one to this gap with equal probability. In this work we ask the following question:

> *Problem 1.* What is the smallest number of queries we need to perform in order to recover g with high probability (up to some unrecoverable global offset)?

Our main contribution is the following result, stated as Theorem 1.

Theorem 1. *There exists a polynomial time algorithm that performs $O(n^{1+o(1)})$ queries, and recovers g (up to some global offset) whp for any $1 - q = \frac{1+\delta}{2}$, where $0 < \delta < 1$ is any positive constant.*

We refer to δ as the *bias*. Notice that when $\delta = 1$, g can be trivially recovered. On the contrary, when $\delta = 0$ exact recovery is impossible. Our result extends our recent work on predicting signed edges [20], and relies on techniques developed there in. Some remarks follow.

Remark 1. The number of queries we perform is $O(n \log n \delta^{-\frac{\log n}{\log \log n}}) = O(n^{1+o(1)})$. We perform queries non-adaptively, specifically we query pairs of nodes uniformly at random. The term $L = \frac{\log n}{\log \log n}$ appears in a natural way, as the diameter of an Erdős-Rényi graph at the connectivity threshold is asymptotically $\sim \frac{\log n}{\log \log n}$ (see Sect. 3).

Remark 2. Observe that even when all $\binom{n}{2}$ possible queries are performed, as long as there is some noise (i.e., $q > 0$), it is not clear a priori whether g can be recovered or not *whp*.

Remark 3. We choose this model for ease of exposition. More generally we can handle queries governed by more general error models, of the form:

$$\tilde{f}(e) = g(x) - g(y) + i \text{ with probability } q_i, 0 \leq i < k.$$

That is, the error does not depend on the group values x and y, but is simply independent and identically distributed over the values 0 to $k - 1$. We outline how our algorithm adapts to this more general case.

Remark 4. In prior work by the authors of this paper [20] a similar model was studied for the case of two latent clusters, i.e., $k = 2$, see also [16]. According to that model, we may query any pair of nodes once, and we receive the correct answer on whether the two nodes are in the same cluster, or not, with probability $1 - q = \frac{1+\delta}{2}$. If we use a model similar to the latter one, it would be difficult to reconstruct the clusters; indeed, even with no errors, a chain of such responses along a path would not generally allow us to determine whether the endpoints of a path were in the same group or not. Our model in this work provides more information and naturally generalizes the two cluster case.

Roadmap. The paper is organized as follows: Sect. 2 presents theoretical preliminaries. Section 3 presents our algorithm, and its analysis. Section 4 surveys related work. Finally, Sect. 5 concludes the paper.

2 Theoretical Preliminaries

We use the following powerful probabilistic results for the proofs in Sect. 3.

Theorem 2 (Chernoff bound, Theorem 2.1 [13]). *Let $X \sim Bin\,(n,p)$, $\mu = np$, $a \geq 0$ and $\varphi(x) = (1+x)\ln(1+x) - x$ (for $x \geq -1$, or ∞ otherwise). Then the following inequalities hold:*

$$\mathbf{Pr}\left[X \leq \mu - a\right] \leq e^{-\mu\varphi\left(\frac{-a}{\mu}\right)} \leq e^{-\frac{a^2}{2\mu}}, \tag{2}$$

$$\mathbf{Pr}\left[X \geq \mu + a\right] \leq e^{-\mu\varphi\left(\frac{-a}{\mu}\right)} \leq e^{-\frac{a^2}{2(\mu+a/3)}}. \tag{3}$$

We define the notion of read-k families, a useful concept when proving concentration results for weakly dependent variables.

Definition 1 (Read-k families). *Let $X_1, \ldots, X_m$ be independent random variables. For $j \in [r]$, let $P_j \subseteq [m]$ and let f_j be a Boolean function of $\{X_i\}_{i \in P_j}$. Assume that $|\{j|i \in P_j\}| \leq k$ for every $i \in [m]$. Then, the random variables $Y_j = f_j(\{X_i\}_{i \in P_j})$ are called a read-k family.*

The following result was proved by Gavinsky et al. for concentration of read-k families. The intuition is that when k is small, we can still obtain strong concentration results.

Theorem 3 (Concentration of Read-k families [10]). *Let $Y_1, \ldots, Y_r$ be a family of read-k indicator variables with $\mathbf{Pr}\,[Y_i = 1] = q$. Then for any $\epsilon > 0$,*

$$\mathbf{Pr}\left[\sum_{i=1}^{r} Y_i \geq (q + \epsilon)r\right] \leq e^{-D_{KL}(q+\epsilon||q)\cdot r/k} \tag{4}$$

and

$$\mathbf{Pr}\left[\sum_{i=1}^{r} Y_i \leq (q - \epsilon)r\right] \leq e^{-D_{KL}(q-\epsilon||q)\cdot r/k}. \tag{5}$$

Here, D_{KL} is Kullback-Leibler divergence defined as

$$D_{\mathrm{KL}}(q||p) = q\log\left(\frac{q}{p}\right) + (1-q)\log\left(\frac{1-q}{1-p}\right).$$

The following corollary of Theorem 3 provides multiplicative Chernoff-type bounds for read-k families. It is derived in a similar way that Chernoff multiplicative bounds are derived from Eqs. (3) and (2), see [17]. Notice that the parameter k appears as an extra factor in denominator of the exponent, that is why when k is relatively small we still obtain meaningful concentration results.

Theorem 4 (Concentration of Read-k families [10]). *Let $Y_1, \ldots, Y_r$ be a family of read-k indicator variables with $\mathbf{Pr}\,[Y_i = 1] = q$. Also, let $Y = \sum_{i=1}^{r} Y_i$. Then for any $\epsilon > 0$,*

$$\mathbf{Pr}\left[Y \geq (1+\epsilon)\mathbb{E}\,[Y]\right] \leq e^{-\frac{\epsilon^2 \mathbb{E}[Y]}{2k(1+\epsilon/3)}} \tag{6}$$

$$\mathbf{Pr}\left[Y \leq (1-\epsilon)\mathbb{E}\,[Y]\right] \leq e^{-\frac{\epsilon^2 \mathbb{E}[Y]}{2k}}. \tag{7}$$

3 Proposed Method

Proof Strategy. Our proposed algorithm is heavily based on our work for the case $k = 2$, a special case of the joint alignment problem of great interest to the social networks' community [20]. In both cases $k = 2$ and $k \geq 3$, the structure of the algorithmic analysis is identical. At a high level, our proof strategy is as follows:

1. We perform $O(n\Delta)$ queries uniformly at random.
2. We compute the probability that a path between x and y provides us with the correct information on $g(x) - g(y)$ or not.
3. We show that there exists a large number of *almost edge-disjoint paths* of length $L = \frac{\log n}{\log \log n}$ between any pair of vertices with probability at least $1 - \frac{1}{n^3}$.
4. To learn the difference $g(x) - g(y)$ for any pair of nodes $\{x, y\}$, we take a majority vote ($k = 2$), or a plurality vote ($k \geq 3$), among the paths we have created. A union bound in combination with (2) shows that *whp* we learn g up to some uknown offset.

Key Differences with Prior Work [20]. While this work relies on [20], there are some key differences. Our main result in [20] is that when there exist two latent clusters ($k = 2$), we can recover them *whp* using $O(n \log n/\delta^4)$ queries, i.e., $\Delta = O(\log n/\delta^4)$. In this work where $k \geq 3$, we set $\Delta = O(\log n\delta^{-L})$, i.e., we perform a larger number of queries. An interesting open question is to reduce the number of queries when $k \geq 3$. Since the models are different, step 2 also differs. Furthermore, the algorithm proposed in [20], and the one we propose here are different; in [20] we use a recursive algorithm that we analyze using Fourier analysis to get a near-optimal result with respect to the number of queries[1]. Here, we use concentration of multivariate polynomials [10], see also [3,14], to analyze the plurality vote of the paths that we construct between a given pair of nodes. Steps 3, 4 are almost identical both in [20], and here. The key difference is that our algorithm requires an average degree $O\left(\frac{\log n}{\delta^L}\right)$ *only for the first level* of certain trees that we grow, for the rest of the levels a branching factor of order $O(\log n)$ suffices.

A Sub-optimal Algorithm for $k = 2$. We describe an algorithm for $k = 2$, that directly generalizes to $k \geq 3$. The caveat is that our proposed algorithm is sub-optimal with respect to the number of queries achieved in [20]. The model for $k = 2$ gets simplified to the following: let $V = [n]$ be the set of n items that belong to two clusters, call them red and blue. Set $g : V \to \{\text{red}, \text{blue}\}$, $R = \{v \in V(G) : g(v) = \text{red}\}$ and $B = \{v \in V(G) : g(v) = \text{blue}\}$, where $0 \leq |R| \leq n$. The function g is unknown and we wish to recover the two clusters R, B by querying pairs of items. (We need not recover the labels, just the clusters.) For each query

---

[1] The information theoretic lower bound on the number of queries is $O(n \log n/\delta^2)$ [11].

Algorithm 1. Learning Joint Alignment for $k = 2$

$L \leftarrow \frac{\log n}{\log \log n}$

Perform $20n \log n \delta^{-L}$ queries uniformly at random.

Let $G(V, E, \tilde{f})$ be the resulting graph, $\tilde{f} : E \rightarrow \{+1, -1\}$

for each item pair x, y **do**

    $\mathcal{P}_{x,y} = \{P_1, \ldots, P_N\} \leftarrow$ Almost-Edge-Disjoint-Paths(x, y)

    $Y_i \leftarrow \prod_{e \in P_i} \tilde{f}(e)$ for $i = 1, \ldots, N$

    $Y_{xy} \leftarrow \sum_{P \in \mathcal{P}_{u,v}} Y_P$

    if $Y_{xy} \geq 0$ **then**

        predict $g(x) = g(y)$

    else

        predict $g(x) \neq g(y)$

    end if

end for

Algorithm 2. Almost-Edge-Disjoint-Paths(x, y)

Require: $G(V, E, \tilde{f})$, $x, y \in V(G)$

    $L \leftarrow \frac{\log n}{\log \log n}$

    $\epsilon \leftarrow \frac{1}{\sqrt{\log \log n}}$

Using Breadth First Search (BFS) grow a tree T_x starting from x as follows. For the first level of the tree, we choose $4 \log n \delta^{-L}$ neighbors of x. For the rest of the tree we use a branching factor equal to $4 \log n$ until it reaches depth equal to ϵL. Similarly, grow a tree T_y rooted at y, node disjoint from T_x of equal depth. From each leaf x_i (y_i) of T_x (T_y) for $i = 1, \ldots, N$ grow node disjoint trees until they reach depth $(\frac{1}{2} + \epsilon)L$ with branching factor $4 \log n$. Finally, find an edge between T_{x_i}, T_{y_i}.

---

we receive the correct answer with probability $1 - q = \frac{1+\delta}{2}$, where $q > 0$ is the corruption probability. That is, for a pair of items x, y such that $g(x) = g(y)$, with probability q it is reported that $g(x) \neq g(y)$, and similarly if $g(x) \neq g(y)$ with probability q it is reported that $g(x) = g(y)$. Since many of the lemmas in this work are proved in a similar way as in [20], we outline the key differences between this work and the proof in [20]. We prove the following Theorem.

Theorem 5. *There exists a polynomial time algorithm that performs* $\Theta(n \log n \delta^{-L})$ *edge queries and recovers the clustering* (R, B) *whp for any gap* $0 < \delta < 1$.

The pseudo-code is shown as Algorithm 1. The algorithm runs over each pair of nodes, and it invokes Algorithm 2 to construct almost edge-disjoint paths for each pair of nodes x, y using Breadth First Search. Note that since we perform $20n \log n \delta^{-L}$ queries uniformly at random, the resulting graph is is asymptotically equivalent to $G \sim G(n, \frac{40 \log n \delta^{-L}}{n})$, see [8, Chap. 1]. Here, $G(n, p)$ is the classic Erdös-Rényi model (a.k.a random binomial graph model) where each possible edge between each pair $(x, y) \in \binom{[n]}{2}$ is included in the graph with probability p independent from every other edge.

It turns out that our algorithm needs an average degree $O\left(\frac{\log n}{\delta^L}\right)$ *only for the first level* of the trees T_x, T_y that we grow from x and y when we invoke Algorithm 2. For all other levels of the grown trees, we need the degree to be only $O(\log n)$. This difference in the branching factors exists in order to ensure that the number of leaves of trees T_x, T_y in Algorithm 2 is amplified by a factor of $\frac{1}{\delta^L}$, which then allows us to apply Theorem 4. Using appropriate data structures, a straight-forward implementation of Algorithm 1 runs in $O(n^2(n+m)) = O(n^3 \log n \delta^{-L})$. Since we use a branching factor of $O(\log n)$ for all except the first two levels of T_x, T_y, we work with the $G(n, p)$ model with $p = \frac{40 \log n}{n}$ to construct the set of almost edge disjoint paths. (Alternatively, one can think that we start with the larger random graph with more edges, and then in the construction of the almost edge disjoint paths we subsample a smaller collection of edges to use in this stage.) The diameter of this graph *whp* grows asymptotically as L [4] for this value of p. We use the $G(n, \frac{40 \log n \delta^{-L}}{n})$ model only in Lemma 1 to prove that every node has degree at least $5 \log n \delta^{-L}$.

Recall that in the case of two clusters $\tilde{f}(e) \in \{-1, +1\}$, indicating whether the oracle answers that the two endpoints of e lie or not in the same cluster. The following result follows by the fact that $\tilde{f}$ agrees with the unknown clustering function g on x, y if the number of corrupted edges along that path P_{xy} is even.

Claim. Consider a path P_{xy} between nodes x, y of length L. Let $R_{xy} = \prod_{e \in P_{xy}} \tilde{f}(e)$. Then,

$$\mathbf{Pr}\left[R_{xy} = 1 | g(x) = g(y)\right] = \mathbf{Pr}\left[R_{xy} = -1 | g(x) \neq g(y)\right] = \frac{1 + (1 - 2q)^L}{2} = \frac{1 + \delta^L}{2}$$

The next lemma is a direct corollary of the lower tail multiplicative Chernoff bound.

Lemma 1. *Let* $G \sim G(n, \frac{40 \log n}{\delta^L n})$ *be a random binomial graph. Then whp all vertices have degree greater than* $5 \log n \delta^{-L}$.

Proof. The degree $deg(x)$ of a node $x \in V(G)$ follows the binomial distribution $Bin(n - 1, \frac{40 \log n}{\delta^L n})$. Set $\gamma = \frac{3}{4}$. Then

$$\mathbf{Pr}\left[deg(x) < 5 \log n \delta^{-L}\right] < e^{-\frac{\gamma^2}{2} 40 \log n \delta^{-L}} \ll n^{-1}.$$

Taking a union bound over n vertices gives the result.

We state the following key lemma, see also [6,9], that shows that we can construct for each pair of nodes x, y a special type of a subgraph $G_{x,y}$.

Lemma 2. *Let* $\epsilon = \frac{1}{\sqrt{\log \log n}}$, *and* $k = \epsilon L$. *For all pairs of vertices* $x, y \in [n]$ *there exists a subgraph* $G_{x,y}(V_{x,y}, E_{x,y})$ *of* G *as shown in Fig. 1, whp. The subgraph consists of two isomorphic vertex disjoint trees* T_x, T_y *rooted at* x, y *each*

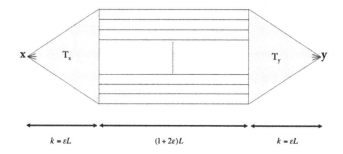

Fig. 1. We create for each pair of nodes x, y two node disjoint trees T_x, T_y whose leaves can be matched via a natural isomorphism and linked with edge disjoint paths. For details, see Lemma 2.

of depth k. T_x and T_y both have a branching factor of $4 \log n \delta^{-L}$ for the first level, and $4 \log n$ for the remaining levels. If the leaves of T_x are $x_1, x_2, \ldots, x_\tau, \tau \geq \delta^{-L} n^{4\epsilon/5}$ then $y_i = f(x_i)$ where f is a natural isomorphism. Between each pair of leaves $(x_i, y_i), i = 1, 2, \ldots, m$ there is a path P_i of length $(1 + 2\epsilon)L$. The paths $P_i, i = 1, 2, \ldots, \tau, \ldots$ are edge disjoint.

We outline that the events hold with large enough probability. For a detailed proof, please check [20]. The only difference with the proof of Lemma 4 in [20] is that for the first level of trees T_x, T_y, we choose $\frac{5 \log n}{\delta^L}$ neighbors of x, y respectively. For all other levels we use a branching factor equal to $4 \log n$. The proof of Theorem 5 follows.

Proof (Theorem 5). Fix a pair of nodes $x, y \in V(G)$, and suppose x, y belong to the same cluster (the other case is treated in the same way). Let $Y_1, \ldots, Y_N$ be the signs of the N edge disjoint paths connecting them, i.e., $Y_i \in \{-1, +1\}$ for all i. Also let $Y = \sum_{i=1}^{N} Y_i$. Notice that $\{Y_1, \ldots, Y_N\}$ is a read-k family where $k = \frac{N}{4 \log n \delta^{-L}}$. By the linearity of expectation, and Lemma 2 we obtain

$$\mathbb{E}[Y] = N \delta^L \geq n^{\frac{4}{5}\epsilon} \delta^L.$$

By applying Theorem 4 we obtain

$$\mathbf{Pr}[Y < 0] = \mathbf{Pr}[Y - \mathbb{E}[Y] < -\mathbb{E}[Y]] \leq \exp\left(-\frac{n^{4/5\epsilon} \delta^L}{\frac{2n^{4/5\epsilon}}{4\delta^{-L} \log n}}\right) = o(n^{-2}).$$

Algorithm for Learning a Joint Alignment, $k \geq 3$. When $q = 0$, so there are no errors from $\tilde{f}(e)$, the edge queries would allow us to determine the difference between the group numbers of vertices at the start and end of any path, and in particular would allow us to determine if the groups were the same. However, when $q > 0$ the actual difference between the cluster ids of x, y, i.e., $g(x) - g(y)$ is perturbed by a certain amount of noise. In the following we discuss how we can tackle this issue. Since the proof of Theorem 1 overlaps with the proof of

---

Algorithm 3. Learning Joint Alignment for $k \geq 3$

$L \leftarrow \frac{\log n}{\log \log n}$

Perform $20n \log n \delta^{-L}$ queries uniformly at random.

Let $G(V, E, \tilde{f})$ be the resulting graph

for each item pair x, y **do**

$\quad \mathcal{P}_{x,y} = \{P_1, \ldots, P_N\} \leftarrow$ Almost-Edge-Disjoint-Paths(x, y)

$\quad Y_i(x, y) \leftarrow \sum_{e \in P_i} \tilde{f}(e)$ for $i = 1, \ldots, N$

$\quad$ Output the plurality vote among $\{Y_1(x, y), \ldots, Y_N(x, y)\}$ as the estimate of $g(x) - g(y)$

end for

Theorem 5 for $k = 2$, we outline the main differences. The idea is still the same: among the differences reported by the large number of paths we create between nodes x, y, the correct answer $g(x) - g(y)$ will be the plurality vote with large enough probability. The pseudocode is shown in Algorithm 3.

Proof (Theorem 1). Let us return to the basic version of our Model, and let $X(e) \in \{-1, 0, 1\}$ for $e = (x, y)$ be

$$\tilde{f}(e) - (g(x) - g(y)) \mod k.$$

Then given a path between two vertices x and y,

$$g(y) = g(x) + \sum_{e \in P_{xy}} \tilde{f}(e) - \sum_{e \in P_{xy}} X(e) \mod k.$$

Our question is now what is $Z_{xy} = \sum_{e \in P_{xy}} X(e) \mod k$. We would like that Z_{xy} be (even slightly) more highly concentrated on 0 than on other values, so that when $g(x) = g(y)$, we find that the sum of the return values from our algorithm, $\sum_{e \in P_{xy}} \tilde{f}(e) \mod k$, is most likely to be 0. We could then conclude by looking over many almost edge-disjoint paths that if this sum is 0 over a plurality of the paths, then x and y are in the same group *whp*, i.e., the plurality value will equal g(y) – g(x) *whp*.

For our simple error model, the sum $\sum_{e \in P_{xy}} X(e) \mod k$ behaves like a simple lazy random walk on the cycle of values modulo k, where the probability of remaining in the same state at each step is q. Let us consider this Markov chain on the values modulo k; we refer to the values as states. Let p_{ij}^t be the probability of going from state i to state j after t steps in such a walk. It is well known than one can derive explicit formulas for p_{ij}^t; see e.g. [7, Chap. XVI.2]. It also follows by simply finding the eigenvalues and eigenvectors of the matrix corresponding to the Markov chain and using that representation. One can check the resulting forms to determine that p_{0j}^t is maximized when $j = 0$, and to determine the corresponding gap $\max_{j \in [1, k-1]} |p_{00}^t - p_{0j}|^t$. Based on this gap, we can apply Chernoff-type bounds as in Theorem 4 to show that the plurality of edge-disjoint paths will have error 0, allowing us to determine whether the endpoints of the path x and y are in the same group with high probability.

The simplest example is with $k = 3$ groups, where we find

$$p_{00}^t = \frac{1}{3} + \frac{2}{3}(1 - 3q/2)^t,$$

and

$$p_{01}^t = p_{02}^t = \frac{1}{3} - \frac{1}{3}(1 - 3q/2)^t.$$

In our case $t = L$, and we see that for any $q < 2/3$, p_{00}^t is large enough that we can detect paths using the same argument as for $k = 2$.

For general k, we use that the eigenvalues of the matrix

$$\begin{bmatrix} 1-q & q/2 & 0 & \cdots & q/2 \\ q/2 & 1-q & q/2 & \cdots & 0 \\ \vdots & \vdots & \vdots & \vdots & \ddots \\ q/2 & 0 & 0 & \cdots & 1-q \end{bmatrix}$$

are $1 - q + q\cos(2\pi j/k)$, $j = 0, \ldots, k-1$, with the j-th corresponding eigenvector being $[1, \omega^j, \omega^{2j}, \ldots, \omega^{j(k-1)}]$ where $\omega = e^{2\pi i/k}$ is a primitive k-th root of unity. Here, i is not an index but the square root of -1, i.e., $i = \sqrt{-1}$. In this case we have

$$p_{00}^t = \frac{1}{k} + \frac{1}{k}\sum_{j=1}^{k-1}(1 - q + q\cos(2\pi j/k))^t.$$

Note that $p_{00}^t > 1/k$. Some algebra reveals that the next largest value of p_{0j}^t belongs to p_{01}^t, and equals

$$p_{01}^t = \frac{1}{k} + \frac{1}{k}\sum_{j=1}^{k-1}\omega^{-j}(1 - q + q\cos(2\pi j/k))^t.$$

We therefore see that the error between ends of a path again have the plurality value 0, with a gap of at least

$$p_{00}^t - p_{01}^t \geq 2(1 - \cos(2\pi/k))(1 - q + q\cos(2\pi/k))^t.$$

This gap is constant for any constant $k \geq 3$ and $q \leq 1/2$.

As we have already mentioned, the same approach could be used for the more general setting where

$$\tilde{f}(e) = g(x) - g(y) + j \text{ with probability } q_j, 0 \leq j < k,$$

but now one works with the Markov chain matrix

$$\begin{bmatrix} q_0 & q_1 & q_2 & \cdots & q_{k-1} \\ q_{k-1} & q_0 & q_1 & \cdots & q_{k-2} \\ \vdots & \vdots & \vdots & \ddots & \vdots \\ q_1 & q_2 & q_3 & \cdots & q_0 \end{bmatrix}.$$

4   Related Work

Many real-world social networks involve both positive and negative interactions or sentiments, that can be positive or negative [15]. The edge sign prediction problem aims to predict the sign $s(x, y) \in \{\pm 1\}$ of an edge $(x, y) \in E(G)$, given the signs of the rest of the edges. Tsourakakis et al. [20] studied this problem both from a theory perspective, using the model proposed in Sect. 3 for $k = 2$ clusters, an empirical perspective, showing that edge-disjoint paths of short length can increase the classification accuracy of the classification algorithms given in [15], especially for pairs of nodes with few common neighbors. A reduction from the planted partition model [1,2,18,19,21], shows that the information theoretic lower bound on the number of queries is $O(n \log n/\delta^2)$, see [2,11]. Mazumdar and Saha [16] study also the problem of clustering using a noisy oracle. When $k = 2$ their model coincides with ours, but when $k \geq 3$ their model is not suitable for learning a joint alignment. For the case of $k = 2$ clusters, they provide a polynomial time algorithm that performs $O(n \log n/\delta^4)$ and runs in $O(n \log n)$ time. For $k \geq 3$, they provide an almost information theoretic optimal algorithm that performs $O(nk \log n/\delta^2)$ queries but does not run in polynomial time, and an algorithm that runs in $O(n \log n + k^6)$ time, but requires $O(k^2 n \log n/\delta^4)$ queries instead. Finally, learning a joint alignment from noisy measurements has several important applications [12,22,23]. Closest to our work lies the work of Chen and Candes who provide stronger theoretical guarantees, using a projected power method to solve the non-convex maximum likelihood estimation problem under our model [5]. Our approach is significantly different, and we conjecture that as in the case of $k = 2$ clusters [20], it may yield asymptotically optimal or near-optimal query complexity.

5   Conclusion

In this work we studied the problem of learning a joint alignment from pairwise differences using a noisy oracle. Based on techniques developed in our previous work [20], we show how we can recover a latent alignment whp using $O(n \log n \delta^{-L})$ queries, where $L = \frac{\log n}{\log \log n}$ is the asymptotic growth of the diameter of an Erdős-Rényi graph at the connectivity threshold. An open question is to improve the dependence on δ. We conjecture, that for constant bias δ, as in the case of $k = 2$ [20], $O(n \log n)$ queries suffice to recover the alignment whp.

References

1. Abbe, E., Bandeira, A.S., Hall, G.: Exact recovery in the stochastic block model. IEEE Trans. Inform. Theory **62**(1), 471–487 (2016)
2. Abbe, E., Sandon, C.: Community detection in general stochastic block models: fundamental limits and efficient algorithms for recovery. In: 2015 IEEE 56th Annual Symposium on Foundations of Computer Science (FOCS), pp. 670–688. IEEE (2015)

3. Alon, N., Spencer, J.H.: The Probabilistic Method. Wiley, New York (2004)
4. Bollobás, B.: Random graphs. In: Modern Graph Theory. Graduate Texts in Mathematics, vol. 184, pp. 215–252. Springer, New York (1998). https://doi.org/10.1007/978-1-4612-0619-4_7
5. Chen, Y., Candes, E.: The projected power method: an efficient algorithm for joint alignment from pairwise differences. arXiv preprint arXiv:1609.05820 (2016)
6. Dudek, A., Frieze, A.M., Tsourakakis, C.E.: Rainbow connection of random regular graphs. SIAM J. Discret. Math. **29**(4), 2255–2266 (2015)
7. Feller, W.: An Introduction to Probability Theory and Its Applications: Volume I, vol. 3. Wiley, New York (1968)
8. Frieze, A., Karoński, M.: Introduction to Random Graphs. Cambridge University Press, Cambridge (2015)
9. Frieze, A., Tsourakakis, C.E.: Rainbow connectivity of sparse random graphs. In: Gupta, A., Jansen, K., Rolim, J., Servedio, R. (eds.) APPROX/RANDOM -2012. LNCS, vol. 7408, pp. 541–552. Springer, Heidelberg (2012). https://doi.org/10.1007/978-3-642-32512-0_46
10. Gavinsky, D., Lovett, S., Saks, M., Srinivasan, S.: A tail bound for read-k families of functions. Random Struct. Algorithms **47**(1), 99–108 (2015)
11. Hajek, B., Wu, Y., Xu, J.: Achieving exact cluster recovery threshold via semidefinite programming. IEEE Trans. Inf. Theory **62**(5), 2788–2797 (2016)
12. Huang, Q.-X., Su, H., Guibas, L.: Fine-grained semi-supervised labeling of large shape collections. ACM Trans. Graph. (TOG) **32**(6), 190 (2013)
13. Janson, S., Luczak, T., Rucinski, A.: Random Graphs, vol. 45. Wiley, Hoboken (2011)
14. Kim, J.H., Vu, V.H.: Concentration of multivariate polynomials and its applications. Combinatorica **20**(3), 417–434 (2000)
15. Leskovec, J., Huttenlocher, D., Kleinberg, J.: Predicting positive and negative links in online social networks. In: Proceedings of the 19th International Conference on World Wide Web (WWW), pp. 641–650. ACM (2010)
16. Mazumdar, A., Saha, B.: Clustering with noisy queries. arXiv preprint arXiv:1706.07510 (2017)
17. McDiarmid, C.: Concentration. In: Habib, M., McDiarmid, C., Ramirez-Alfonsin, J., Reed, B. (eds.) Probabilistic Methods for Algorithmic Discrete Mathematics. Algorithms and Combinatorics, vol. 16, pp. 195–248. Springer, Heidelberg (1998). https://doi.org/10.1007/978-3-662-12788-9_6
18. McSherry, F.: Spectral partitioning of random graphs. In: Proceedings of the 42nd IEEE Symposium on Foundations of Computer Science (FOCS), pp. 529–537. IEEE (2001)
19. Perry, A., Wein, A.S.: A semidefinite program for unbalanced multisection in the stochastic block model. In: 2017 International Conference on Sampling Theory and Applications (SampTA), pp. 64–67. IEEE (2017)
20. Tsourakakis, C.E., Mitzenmacher, M., Błasiok, J., Lawson, B., Nakkiran, P., Nakos, V.: Predicting positive and negative links with noisy queries: theory & practice. arXiv preprint arXiv:1709.07308 (2017)
21. Vu, V.: A simple SVD algorithm for finding hidden partitions. arXiv preprint arXiv:1404.3918 (2014)
22. Wang, L., Singer, A.: Exact and stable recovery of rotations for robust synchronization. Inf. Infer. J. IMA **2**(2), 145–193 (2013)
23. Zach, C., Klopschitz, M., Pollefeys, M.: Disambiguating visual relations using loop constraints. In: 2010 IEEE Conference on Computer Vision and Pattern Recognition (CVPR), pp. 1426–1433. IEEE (2010)

Analysis of Relaxation Time in Random Walk with Jumps

Konstantin Avrachenkov[1]([✉]) and Ilya Bogdanov[2]

[1] Inria Sophia Antipolis, Valbonne, France
k.avrachenkov@inria.fr
[2] Higher School of Economics, Moscow, Russia
ilya160897@gmail.com

Abstract. We study the relaxation time in the random walk with jumps. The random walk with jumps combines random walk based sampling with uniform node sampling and improves the performance of network analysis and learning tasks. We derive various conditions under which the relaxation time decreases with the introduction of jumps.

Keywords: Random walk on graph · Random walk with jumps
Relaxation time · Spectral gap · Network sampling · Network analysis
Learning on graphs

1 Introduction

In the present work we study the relaxation time or equivalently the spectral gap of the random walk with jumps on a general weighted undirected graph. The random walk with jumps can be viewed as a random walk on a combination of the original graph and the complete graph weighted by a scaled parameter. This parameter determines the rate of jumps from the current node to an arbitrary node. The random walk with jumps has similarities with PageRank [18]. In fact, it coincides with PageRank on the regular graphs but differs on the irregular graphs. In the case of the random walk with jumps, the jump probability depends on the node degree. The higher the degree of the current node is, the less likely the random walk will jump out of the node to an arbitrary node. The random walk with jumps can also be viewed as a particular case of the generalisation of PageRank with node-dependent restart [6].

The random walk with jumps has been introduced in [5] to improve the random walk based network sampling or respondent driven sampling [21] by combining the standard random walk based sampling with uniform node sampling. A big advantage of the random walk with jumps in comparison with PageRank is that, as opposite to PageRank, the random walk with jumps on an undirected graph is a time-reversible process and its stationary distribution is available in a simple, explicit form. In particular, this allows us to unbias efficiently the random walk, which is node degree biased on irregular graphs. This comes with a

© Springer International Publishing AG, part of Springer Nature 2018
A. Bonato et al. (Eds.): WAW 2018, LNCS 10836, pp. 70–82, 2018.
https://doi.org/10.1007/978-3-319-92871-5_6

price. The price is the difficulty to control the relaxation time. In the case of PageRank the relaxation time is bounded by the reciprocal of the restart probability [10,15]. In the case of the random walk with jumps, there is no simple connection with the jump parameter. In [5] the authors have shown that under a natural condition on the clustering structure of the graph, the relaxation time decreases with the increase of the jump parameter.

Let us mention a few more applications of the random walk with jumps beyond network sampling [5,17,19]. The random walk with jumps has been used in the context of graph-based semi-supervised learning [2,14]. The random walk with jumps has also been used as a main building block in the quick algorithm for finding largest degree nodes in a graph [4]. A continuous-time version of the random walk with jumps has an application in epidemiology [12].

All this motivates us to take another look at the relaxation time of the random walk with jumps. In particular, we are now able to give a necessary and sufficient condition for the improvement of the relaxation time on weighted graphs. We give an example showing that there are weighted graphs where introducing jumps increases the relaxation time. The necessary and sufficient conditions are not easy to interpret. Therefore we derive a series of simpler sufficient conditions. One new sufficient condition, similar in spirit to the condition in [5], indicates that on graphs with clusters, the relaxation time improves with the introduction of jumps. The other new sufficient conditions require the spectral gap of the original graph to be smaller than the reciprocal of the squared coefficient of variation of the nodes' degrees, thus establishing a connection between the measure of graph irregularity and the relaxation time. We expect that the derived conditions are satisfied in most complex networks (either due to clustering structure or due to small spectral gap). Thus, the present study confirms that it is safe and in most cases beneficial to use the random walk with jumps for complex network analysis.

The structure of the paper is as follows: in the next section we define the random walk with jumps and provide necessary background material. In Sect. 3 we discuss the application of Dobrushin coefficient for large jump rates. Then, in Sect. 4 we discuss necessary and sufficient conditions for the reduction of the relaxation time when the jump rate is small. In Sect. 5 we provide a series of sufficient conditions, which are easier to interpret and to verify. In particular, we provide a sufficient condition in terms of the coefficient of variation of nodes' degrees. In Sect. 6 we give interesting numerical illustrations. We conclude the paper with a conjecture in Sect. 7.

2   Definitions and Preliminaries

Most of the analysis in the present article is for the case of a general weighted undirected graph G with vertex set $V(G)$, $|V(G)| = n$, and edge set $E(G)$, $|E| = m$, defined by the *weighted adjacency matrix* $A = (a_{ij})$ with elements

$$a_{ij} = \begin{cases} \text{weight of edge } (i,j), \text{ if } i \sim j, \\ 0, \qquad\qquad\qquad\qquad \text{otherwise.} \end{cases}$$

Unless stated otherwise, we assume that G is connected.

Denote by $\underline{1}$ the column vector of ones of appropriate dimension. Then, $d = A\underline{1}$ is the vector of weighted degrees of vertices and $D = Diag(d)$ is the diagonal matrix with vector d on the main diagonal.

The Standard Random Walk (SRW) is a discrete-time Markov chain $\{X_t, t = 0, 1, ...\}$ on the vertex set $V(G)$ with the transition probability matrix

$$P = D^{-1}A, \tag{1}$$

whose elements are

$$p_{ij} = P[X_{t+1} = j | X_t = i] = \begin{cases} w_{ij}/d_i, & \text{if } i \sim j, \\ 0, & \text{otherwise.} \end{cases}$$

SRW is a time-reversible process with the stationary distribution

$$\pi_i = \frac{d_i}{2m}, \quad i = 1, ..., n. \tag{2}$$

Since SRW is time-reversible, its transition matrix is similar to a symmetric matrix and hence the eigenvalues of the transition matrix are real, semi-simple and can be indexed as follows:

$$1 = \lambda_1 \geq \lambda_2 \geq ... \geq \lambda_n \geq -1.$$

Denote λ_* the maximum modulus eigenvalue of P different from $+1$ and -1 and call $\gamma(P) = 1 - |\lambda_*|$ the spectral gap.

The relaxation time t_{rel} is then defined by

$$t_{rel} = \frac{1}{\gamma(P)}. \tag{3}$$

One interpretation of the relaxation time is as follows [16]: if $t \geq t_{rel}$, then the standard deviation of $P^t f$ is bounded by $1/e$ times the standard deviation of f. Also, the following inequality (see, e.g., [1,9,16]) indicates a strong relation between the relaxation time and mixing time

$$(\log(1/\varepsilon) + \log(1/2))(t_{rel} - 1) \leq t_{mix}(\varepsilon) \leq (\log(1/\varepsilon) + \log(1/\pi_{min}))t_{rel}.$$

In particular, the above inequality suggests that for a finite Markov chain and for small enough ε, the ε-mixing time is very close to $\log(1/\varepsilon)t_{rel}$.

Now let us define the random walk with jumps (RWJ). It is a random walk on a combination of the original graph and the complete graph weighted by a scaled parameter α/n [5]. Specifically, let us modify the adjacency matrix in the following way

$$A(\alpha) = A + \frac{\alpha}{n}\underline{1}\,\underline{1}^T. \tag{4}$$

Note that the new degree matrix is given by $D(\alpha) = D + \alpha I$, where I is the identity matrix. Then, the random walk on the modified graph is described by the following transition probability matrix

$$P(\alpha) = D^{-1}(\alpha)A(\alpha), \tag{5}$$

with elements

$$p_{ij}(\alpha) = \begin{cases} \frac{w_{ij}+\alpha/n}{d_i+\alpha}, & \text{if } i \sim j, \\ \frac{\alpha/n}{d_i+\alpha}, & \text{otherwise,} \end{cases}$$

Since RWJ is again a random walk on a weighted undirected graph, it is time-reversible Markov chain with semi-simple eigenvalues. The stationary distribution of RWJ also has a simple form

$$\pi_i(\alpha) = \frac{d_i + \alpha}{2m + \alpha n}, \quad i = 1, ..., n. \tag{6}$$

The modified transition matrix $P(\alpha)$ can be rewritten as follows:

$$P(\alpha) = (D + \alpha I)^{-1} D P + (D + \alpha I)^{-1} \alpha \mathbf{1} \left(\frac{1}{n}\mathbf{1}^T\right). \tag{7}$$

We note that if the graph is regular, i.e., $D = dI$, the above expression reduces to

$$P(\alpha) = \frac{d}{d+\alpha} P + \frac{\alpha}{d+\alpha} \mathbf{1} \left(\frac{1}{n}\mathbf{1}^T\right),$$

which is the transition matrix for PageRank (PR) [18] with the damping factor $d/(d+\alpha)$. Thus, in the case of a regular graph RWJ coincides with PR. However, when the graph has inhomogeneous degrees, these two concepts are different. From the equivalence of RWJ to PR on the regular graph, we can immediately conclude that the relaxation time is monotonously decreasing with α when the graph is regular. When the graph is irregular, the situation becomes much more complex.

We also would like to note that RWJ can be viewed as a node-dependent PageRank [6], where the restart probability at each node is given by $\alpha/(d_i + \alpha)$. Thus, in contrast to PR, RWJ restarts with smaller probabilities from higher degree nodes.

3 Application of Dobrushin Coefficient

The Dobrushin ergodic coefficient (see, e.g., [8,11,20]) can be used to obtain a lower bound on the spectral gap $\gamma(P)$ of a Markov chain. The Dobrushin ergodic coefficient is given by

$$\delta(P) = \frac{1}{2} \max_{i,j \in V} \sum_{k \in V} |p_{ik} - p_{jk}|, \tag{8}$$

or, equivalently,

$$\delta(P) = 1 - \min_{i,j \in V} \sum_{k \in V} p_{ik} \wedge p_{jk}. \tag{9}$$

In the case of RWJ, we can obtain a simple upper bound on $\delta(P)$ by taking the smallest element in the transition probability matrix. Namely, we obtain

$$\delta(P) \leq 1 - \min_{i,j \in V} \sum_{k \in V} \frac{\alpha/n}{d_{max} + \alpha} = 1 - \frac{\alpha}{d_{max} + \alpha}, \tag{10}$$

where d_{max} is the maximal degree in the graph. Since $\gamma(P) \geq 1 - \delta(P)$, we also obtain a lower bound for the spectral gap

$$\gamma(P(\alpha)) \geq \frac{\alpha}{d_{max} + \alpha}. \tag{11}$$

And since $\alpha/(d_{max} + \alpha) \to 1$ as $\alpha \to \infty$, we have

Proposition 1. *For any undirected graph G, there always exists $\bar{\alpha} = \bar{\alpha}(G)$ such that for all $\alpha > \bar{\alpha}$, we have $\gamma(P(\alpha)) > \gamma(P)$.*

For the regular graphs with degree d, the bound $\alpha/(d + \alpha)$ is quite tight. In fact, for the regular graphs, as was noted at the end of the previous section $\frac{\alpha}{d+\alpha}$ corresponds to the restart probability and the exact value of the spectral gap is $\frac{d}{d+\alpha}\lambda_*(P)$ [10, 15]. However, we have observed that for irregular graphs the bound (11) can be very loose.

4   Conditions in the Case of Small Jump Rate

Let us analyse in this section the effect of small jump rate on the relaxation time.

Denote by $v_*(\alpha)$ the eigenvector corresponding to $\lambda_*(\alpha)$, that is

$$P(\alpha)v_*(\alpha) = \lambda_*(\alpha)v_*(\alpha). \tag{12}$$

For brevity, we shall write $v_* = v_*(0)$ and $\lambda_* = \lambda_*(0)$. We also need the following preliminary result

Lemma 1. *The eigenelements $v_*(\alpha)$ and $\lambda_*(\alpha)$ are analytic functions with respect to α.*

Proof: It is known from [13, Chap. 2] (see also [3, 7]) that if the eigenvalues of the perturbed matrix are semi-simple, then the eigenvalues as well as eigenvectors can be expanded as power series with positive integer powers. Since in our case, RWJ is time-reversible, its eigenvalues are semi-simple and the statement of the lemma follows.   □

Next, we are in a position to provide necessary and sufficient conditions for the improvement of the relaxation time for small jump rates.

Theorem 1. *For sufficiently small α, in the case $\lambda_* < 0$, the spectral gap increases, and equivalently, the relaxation time decreases with respect to α.*

If $\lambda_* > 0$, for sufficiently small α the necessary and sufficient condition for the decrease of the relaxation time is

$$\frac{1}{n}(\underline{1}^T v_*)^2 < \lambda_* v_*^T v_*. \tag{13}$$

In addition, we have the following asymptotics:

$$\lambda_*(\alpha) = \lambda_*(0) + \alpha \frac{\frac{1}{n}(v_*^{(0)})^T \underline{11}^T v_*^{(0)} - \lambda_*^{(0)}(v_*^{(0)})^T v_*^{(0)}}{(v_*^{(0)})^T D v_*^{(0)}} + O(\alpha^2).$$

Proof: According to Lemma 1, the eigenelements $\lambda_*(\alpha)$ and $v_*(\alpha)$ can be expanded as power series

$$\lambda_*(\alpha) = \lambda_*^{(0)} + \alpha\lambda_*^{(1)} + O(\alpha^2), \tag{14}$$

$$v_*(\alpha) = v_*^{(0)} + \alpha v_*^{(1)} + O(\alpha^2), \tag{15}$$

for sufficiently small α. If we set $\alpha = 0$, we obtain $\lambda_*^{(0)} = \lambda_*(0) = \lambda_*$ and $v_*^{(0)} = v_*(0) = v_*$ such that

$$D^{-1}Av_* = \lambda_* v_*,$$

or equivalently,

$$Av_* = \lambda_* Dv_*. \tag{16}$$

It is also convenient to rewrite (12) as a generalized eigenvalue problem

$$(A + \frac{\alpha}{n}\underline{1}\underline{1}^T)v_*(\alpha) = \lambda_*(\alpha)(D + \alpha I)v_*(\alpha). \tag{17}$$

Substituting the power series (14) and (15) into Eq. (17) and equating coefficients in α-terms, i.e.,

$$(A + \frac{\alpha}{n}\underline{1}\underline{1}^T)(v_*^{(0)} + \alpha v_*^{(1)} + O(\alpha^2)) =$$

$$(\lambda_*^{(0)} + \alpha\lambda_*^{(1)} + O(\alpha^2))(D + \alpha I)(v_*^{(0)} + \alpha v_*^{(1)} + O(\alpha^2)),$$

yields

$$\frac{1}{n}\underline{1}\underline{1}^T v_*^{(0)} + Av_*^{(1)} = \lambda_*^{(0)}Iv_*^{(0)} + \lambda_*^{(1)}Dv_*^{(0)} + \lambda_*^{(0)}Dv_*^{(1)}. \tag{18}$$

Now let us multiply the above equation by v_*^T from the left to obtain

$$\frac{1}{n}(v_*^{(0)})^T\underline{1}\underline{1}^T v_*^{(0)} + (v_*^{(0)})^T Av_*^{(1)} \tag{19}$$

$$= (v_*^{(0)})^T\lambda_*^{(0)}v_*^{(0)} + (v_*^{(0)})^T\lambda_*^{(1)}Dv_*^{(0)} + (v_*^{(0)})^T\lambda_*^{(0)}Dv_*^{(1)}.$$

Due to symmetry, the Eq. (16) can be rewritten as

$$v_*^T A = \lambda_* v_*^T D,$$

and hence

$$(v_*^{(0)})^T Av_*^{(1)} = (v_*^{(0)})^T\lambda_*^{(0)}Dv_*^{(1)},$$

which simplifies (19) to

$$\frac{(v_*^{(0)})^T\underline{1}\underline{1}^T v_*^{(0)}}{n} = (v_*^{(0)})^T\lambda_*^{(0)}v_*^{(0)} + (v_*^{(0)})^T\lambda_*^{(1)}Dv_*^{(0)}.$$

Thus, $\lambda_*^{(1)}$ can be expressed as

$$\lambda_*^{(1)} = \frac{\frac{1}{n}(v_*^{(0)})^T\underline{1}\underline{1}^T v_*^{(0)} - \lambda_*^{(0)}(v_*^{(0)})^T v_*^{(0)}}{(v_*^{(0)})^T Dv_*^{(0)}}. \tag{20}$$

Clearly, the denominator in (20) is always positive. Now, consider two cases:

Case 1. $\lambda_* < 0$

In this case the numerator in (20) is always positive. Then $\lambda_*^{(1)}$ is also positive. By expansion (14), when α is sufficiently small, the absolute value of λ_* is decreasing with respect to α.

Case 2. $\lambda_* > 0$

Again, by expansion (14), for sufficiently small α the value of $\lambda_*(\alpha)$ decreases in α if and only if $\lambda_*^{(1)}$ is negative, i.e., when the numerator is negative. This is precisely what was stated in the theorem's condition. $\qquad\square$

5   Sufficient Conditions with Easier Interpretation

Even though the condition in Theorem 1 is necessary and sufficient, it is not easy to use and does not have an easy intuitive interpretation. Next, we shall derive a series of sufficient conditions with easier interpretation and verification. Towards this goal, let us transform the condition in Theorem 1 to an equivalent form using the combinatorial Laplacian. Specifically, we shall use the combinatorial Laplacian of the complete graph:

$$L_K = nI - \underline{1}\,\underline{1}^T.$$

Lemma 2. *The condition (13) is equivalent to*

$$1 - \lambda_* < \frac{v_*^T L_K v_*}{n v_*^T v_*},$$

and if $\lambda_* > 0$, the condition (13) is equivalent to

$$\gamma(P) < \frac{\sum_{i,j}(v_{*i} - v_{*j})^2}{n \sum_i v_{*i}^2}. \tag{21}$$

Proof: Using the definition of the Laplacian, we can write

$$\frac{1}{n}(\underline{1}^T v_*)^2 = \frac{1}{n}v_*^T \underline{1}\,\underline{1}^T v_* = \frac{1}{n}v_*^T(nI - L_K)v_* = v_*^T v_* - \frac{1}{n}v_*^T L_K v_*.$$

Thus, we can rewrite condition (13) in a new form:

$$1 - \lambda_* < \frac{v_*^T L_K v_*}{n v_*^T v_*}.$$

The other equivalent form follows immediately from the definitions of the spectral gap and L_K. $\qquad\square$

Next we provide a couple of sufficient conditions with easy interpretation in the context of complex networks.

Corollary 1. *If*

$$\gamma(P) < \frac{1}{n}, \tag{22}$$

then, for sufficiently small $\alpha > 0$, the spectral gap of $P(\alpha)$ is larger than the spectral gap of P.

Proof: Clearly, $\sum_{i,j}(v_{*i}(\alpha) - v_{*j}(\alpha))^2 > \sum_i v_{*i}(\alpha)^2$. Because $v_*(\alpha) \perp \pi(\alpha)$ and so there are both positive and negative numbers among $v_{*i}(\alpha)$. Each number $v_{*i}(\alpha)$ has at least one number of opposite sign $v_{*j}(\alpha)$ and such that $(v_{*i}(\alpha) - v_{*j}(\alpha))^2 > v_{*i}(\alpha)^2$. □

The above corollary has the following simple and useful interpretation: If P has a sufficiently small gap, the addition of jumps with small rate always improves the relaxation time. It is known that many complex networks have very small spectral gap and thus this corollary gives an explanation why the relaxation time typically improves with the addition of jumps in complex networks. The next corollary refines the above argument.

Corollary 2. *Denote the proportion of negative and positive $v_{*i}(\alpha)$ as $\mu : 1 - \mu$. Then, for sufficiently small α, if*

$$\gamma(P) < \min(\mu, 1 - \mu), \tag{23}$$

the relaxation time decreases with respect to α.

Proof: For each positive number $v_{*i}(\alpha)$ has μn numbers of opposite sign $v_{*j}(\alpha)$ such that $(v_{*i}(\alpha) - v_{*j}(\alpha))^2 > v_{*i}(\alpha)^2$. Analogously, each negative number $v_{*i}(\alpha)$ has $(1 - \mu)n$ numbers of opposite sign $v_{*j}(\alpha)$ and for them $(v_{*i}(\alpha) - v_{*j}(\alpha))^2 > v_{*i}(\alpha)^2$. So, $\sum_{i,j}(v_{*i}(\alpha) - v_{*j}(\alpha))^2 \geq n \min(\mu, 1 - \mu) \sum_i v_{*i}(\alpha)^2$. □

As a result of Corollary 2, the closer μ is to $1/2$, the better. Often complex networks have clustering structure. The eigenvector v_* can be interpreted as a variant of the Fiedler vector. Thus, if a complex network can be divided into two clusters of similar sizes, the value of μ will be close to $1/2$ or at least far from zero. In such a case, the spectral gap of the original transition matrix P does not need to be small for the addition of jumps to improve the relaxation time. The above statement is similar in spirit to the condition given in [5].

Let us now provide a sufficient condition in terms of nodes' degrees.

Theorem 2. *Let the following condition hold:*

$$\gamma(P) < 4\frac{(d_1 + d_2 + \cdots + d_n)^2}{n(d_1^2 + d_2^2 + \cdots + d_n^2)}. \tag{24}$$

Then, if $\alpha > 0$ is sufficiently small, $\gamma(P(\alpha)) > \gamma(P)$.

Proof: We replace the condition (21) with the following more stringent condition:

$$\gamma(P) < \min_{f \perp \pi} \frac{\sum_{i \sim j}(f_i - f_j)^2}{n \sum f_i^2}. \tag{25}$$

Let us find the minimum of the RHS of the above expression over all vectors f, orthogonal to $\pi(\alpha)$, i.e., for vectors satisfying $\sum f_i d_i = 0$. So it will be enough for $\gamma(P)$ to be less than the RHS of the condition (25).

Note that the RHS of (25) is homogeneous in f, so without loss of generality we can set $\sum f_i^2 = 1$.

We shall use the method of Lagrange multipliers for the objective function $\sum_{i,j}(f_i - f_j)^2$ and the following constraints:

$$\sum_{i=1}^{n} f_i^2 = 1, \tag{26}$$

$$\sum_{i=1}^{n} f_i d_i = 0. \tag{27}$$

The expression $\sum_{i,j}(f_i - f_j)^2$ taking into account the latter constraint can be rewritten as $2n - 2(f_1 + f_2 + \cdots + f_n)^2$. Therefore we are looking for the maximum of $(f_1 + f_2 + \cdots + f_n)^2$, which will give us the minimum of $\sum_{i,j}(f_i - f_j)^2$. Let us find the maximal absolute values of the function $(f_1 + f_2 + \cdots + f_n)$ with respect to the above constraints. Obviously, these extreme values exist (sums cannot be more than n and less than $-n$). Furthermore, their absolute values coincide, because domains of these two sums is symmetric with the center in the origin. So the set of values of $(f_1 + f_2 + \cdots + f_n)$ is centrally symmetric and the maximal absolute values of $(f_1 + f_2 + \cdots + f_n)$ and $-(f_1 + f_2 + \cdots + f_n)$ are the same.

Looking at the geometry of our maximization problem, we can see that the constraints give us a sphere cut by the hyperplane $\sum f_i d_i = 0$. We are looking for the touching points of this set and some hyperplane $(f_1 + f_2 + \cdots + f_n) = c$. Clearly, there will be exactly two centrally-symmetric points of touching, giving us the maxima of the absolute values (the domain of this sum is symmetric with the center of the symmetry in the origin, so the set of values of $(f_1 + f_2 + \cdots + f_n)$ is centrally symmetric).

Now, consider the Lagrange function:

$$L = (f_1 + f_2 + \cdots + f_n) - \lambda_1 \left(\sum_i f_i^2 - 1 \right) - \lambda_2 \sum f_i d_i. \tag{28}$$

Taking partial derivatives with respect to f_i for $i = 1, ..., n$, we obtain

$$0 = 1 - 2\lambda_1 f_i - \lambda_2 d_i. \tag{29}$$

Firstly, multiply all these equations with d_i, respectively, and sum all of them (here we also use Eq. (27)). The result is

$$0 = 2m - \lambda_2 \sum_i d_i^2. \tag{30}$$

Thus, $\lambda_2 = \frac{2m}{\sum_i d_i^2}$, where m is the total number of edges.

Secondly, multiply all equations with f_i, respectively, and sum all of them (here we additionally use Eqs. (26) and (27)). The result is

$$0 = S - 2\lambda_1, \tag{31}$$

where $S = (f_1 + f_2 + \cdots + f_n)$. Thus, $\lambda_1 = S/2$.

And finally, just sum all equations (here we also use Eqs. (26) and (27)) to get

$$0 = n - 2\lambda_1 S - \lambda_2 (2m). \tag{32}$$

If we know λ_1 and λ_2, we also know $S^2 = n - \frac{8m^2}{\sum_i d_i^2}$. Substituting this into our initial expression that we want to minimize, we obtain

$$\min_{f \perp \pi} \frac{\frac{1}{n} \sum_{i,j} (f_i - f_j)^2}{\sum f_i^2} = \frac{1}{n} \left(2n - 2 \left(n - \frac{8m^2}{\sum_i d_i^2} \right) \right) = 4 \frac{(d_1 + d_2 + \cdots + d_n)^2}{n(d_1^2 + d_2^2 + \cdots + d_n^2)}. \tag{33}$$

Thus, if inequality (25) holds, the condition of Theorem 1 also holds, and hence $\gamma(P(\alpha)) > \gamma(P)$ for sufficiently small α. $\qquad\square$

We can also rewrite the condition of the above theorem as follows:

Corollary 3. *Let the following condition hold:*

$$\gamma(P) < 4 \frac{(\bar{d})^2}{\overline{d^2}}, \tag{34}$$

where

$$\bar{d} = \frac{d_1 + d_2 + \cdots + d_n}{n},$$

and

$$\overline{d^2} = \frac{d_1^2 + d_2^2 + \cdots + d_n^2}{n}.$$

Then, if $\alpha > 0$ is sufficiently small, $\gamma(P(\alpha)) > \gamma(P)$.

The quantity $(\bar{d})^2/\overline{d^2}$ is the reciprocal of the efficiency or of the squared coefficient of variation. It is also sometimes referred to as the signal-to-noise ratio. From Corollary 3 we can see that the more "irregular" the degree sequence is, the more stringent is the condition on the spectral gap.

The following simple sufficient condition using only $\bar{d}$ and d_{max} also holds.

Corollary 4. *For sufficiently small α, $\gamma(P(\alpha)) > \gamma(P)$ is true for all graphs such that*

$$\gamma(P) < 4\bar{d}/d_{max}. \tag{35}$$

Proof: The RHS of the inequality (24) is greater than $\bar{d}/d_{max}$. $\qquad\square$

6   Examples

Let us first demonstrate in this section that there exist weighted graphs for which the introduction of jumps actually increases the relaxation time. Towards this end, we consider a weighted graph of size 2 with the following adjacency matrix:

$$A = \begin{bmatrix} a_{11} & a_{12} \\ a_{12} & a_{22} \end{bmatrix}.$$

The characteristic equation for the generalized eigenvalue problem $Av = \lambda Dv$ takes the form

$$\det(A - \lambda D) = \det \begin{bmatrix} a_{11} - \lambda(a_{11} + a_{12}) & a_{12} \\ a_{12} & a_{22} - \lambda(a_{22} + a_{12}) \end{bmatrix} = 0.$$

It has two solutions

$$\lambda_1 = 1 \quad \text{and} \quad \lambda_* = \frac{\det(A)}{(a_{11} + a_{12})(a_{22} + a_{12})}.$$

The eigenvector corresponding to λ_* is given by

$$v_* = \begin{bmatrix} 1 \\ -\frac{a_{11}+a_{12}}{a_{22}+a_{12}} \end{bmatrix} C.$$

Let us calculate the numerator in the expression (20)

$$\frac{1}{2}(v_*^T 1)^2 - \lambda_* v_*^T v_* =$$

$$\frac{(a_{22} - a_{11})^2 (a_{11} + a_{12})(a_{22} + a_{12}) - 2\det(A)[(a_{11} + a_{12})^2 + (a_{22} + a_{12})^2]}{2(a_{11} + a_{12})(a_{22} + a_{12})^3}.$$

Now if we choose the weights such that $\det(A)$ is close to zero or just zero and $a_{11} \neq a_{22}$, the spectral gap will decrease, and consequently the relaxation time will increase when α increases from zero. Thus, in general, the conditions of Theorem 1 need to be checked in the case of weighted graphs if one wants to be sure that the relaxation time decreases with the increase of the jump rate.

We also tried to construct an example of unweighted graph when the relaxation time increases with the increase in the jump rate. Somewhat surprisingly, it appears to be hard to construct such an example. In fact, we have checked all non-isomorphic graphs of sizes up to $n \leq 9$ from the Brendan McKay's collection [22] and could not find a single example when the relaxation time increases with the introduction of small jump rate. The code of this verification can be found at [23].

We have also checked various random graph models like Erdős-Rényi graph or non-homogeneous Stochastic Block Model and could not find an example of an unweighted graph for which the relaxation time increases with the introduction of jumps.

7   Conclusion and Future Research

We have analysed the spectral gap, or equivalently the relaxation time, in the random walk with jumps. We have obtained a necessary and sufficient condition for the decrease of the relaxation time when the jump rate increases from zero. These conditions unfortunately do not have easy interpretation. Therefore, we have proceeded with the derivation of several sufficient conditions with easy interpretation. Some of these sufficient conditions can also be easily verified. The derived sufficient conditions show that in most complex networks the relaxation time should decrease with the introduction of jumps. We have also demonstrated that there exist weighted graphs for which the relaxation time increases with the introduction of jumps. On the other hand, we could not find such an example in the case of unweighted graphs. At the moment, we tend to conjecture that the introduction of jumps always improves the relaxation time in the case of unweighted graphs.

Acknowledgements. This work was partially supported by the joint laboratory Nokia Bell Labs - Inria.

References

1. Aldous, D., Fill, J.: Reversible Markov chains and random walks on graphs. Monograph in Preparation (2002). http://www.stat.berkeley.edu/~aldous/RWG/book.html
2. Avrachenkov, K., Chebotarev, P., Mishenin, A.: Semi-supervised learning with regularized Laplacian. Optim. Methods Softw. **32**(2), 222–236 (2017)
3. Avrachenkov, K., Filar J.A., Howlett P.G.: Analytic Perturbation Theory and Its Application. SIAM (2013)
4. Avrachenkov, K., Litvak, N., Sokol, M., Towsley, D.: Quick detection of nodes with large degrees. Internet Math. **10**(1–2), 1–19 (2014)
5. Avrachenkov, K., Ribeiro, B., Towsley, D.: Improving random walk estimation accuracy with uniform restarts. In: Kumar, R., Sivakumar, D. (eds.) WAW 2010. LNCS, vol. 6516, pp. 98–109. Springer, Heidelberg (2010). https://doi.org/10.1007/978-3-642-18009-5_10
6. Avrachenkov, K., van der Hofstad, R., Sokol, M.: Personalized pagerank with node-dependent restart. In: Bonato, A., Graham, F.C., Prałat, P. (eds.) WAW 2014. LNCS, vol. 8882, pp. 23–33. Springer, Cham (2014). https://doi.org/10.1007/978-3-319-13123-8_3
7. Baumgärtel, H.: Analytic Perturbation Theory for Matrices and Operators. Birkhäuser, Basel (1985)
8. Brémaud, P.: Markov Chains: Gibbs Fields, Monte Carlo Simulation, and Queues, vol. 31. Springer, New York (1999). https://doi.org/10.1007/978-1-4757-3124-8
9. Chung, F.: Spectral graph theory. American Math. Soc. (1997)
10. Haveliwala, T., Kamvar, S.: The second eigenvalue of the Google matrix. Stanford Technical Report (2003)
11. Ipsen, I.C.F., Selee, T.M.: Ergodicity coefficients defined by vector norms. SIAM J. Matrix Anal. Appl. **32**(1), 153–200 (2011)

12. Jacobsen, K.A., Tien, J.H.: A generalized inverse for graphs with absorption. Linear Algebra Appl. **537**, 118–147 (2018)
13. Kato, T.: Perturbation Theory for Linear Operators, vol. 132. Springer, Heidelberg (1995). https://doi.org/10.1007/978-3-642-66282-9
14. Kondor, R.I., Lafferty, J.: Diffusion kernels on graphs and other discrete input spaces. In: Proceedings of ICML, pp. 315–322 (2002)
15. Langville, A.N., Meyer, C.D.: Google's PageRank and Beyond: The Science of Search Engine Rankings. Princeton University Press, Princeton (2006)
16. Levin, D.A., Peres, Y., Wilmer, E.L.: Markov Chains and Mixing Times. American Math. Soc. (2008)
17. Murai, F., Ribeiro, B., Towsley, D. and Wang P.: Characterizing directed and undirected networks via multidimensional walks with jumps. ArXiv preprint arXiv:1703.08252 (2017)
18. Page, L., Brin, S., Motwani, R., Winograd, T.: The PageRank citation ranking: Bringing order to the web. Stanford InfoLab Research Report (1999)
19. Ribeiro, B., Towsley, D.: Estimating and sampling graphs with multidimensional random walks. In: Proceedings of the 10th ACM SIGCOMM Conference on Internet Measurement, pp. 390–403 (2010)
20. Seneta, E.: Non-Negative Matrices and Markov Chains, Revised Printing edition. Springer, New York (2006). https://doi.org/10.1007/0-387-32792-4
21. Volz, E., Heckathorn, D.D.: Probability based estimation theory for respondent driven sampling. J. Official Stat. **24**(1), 79 (2008)
22. Brendan McKay's graph collection. http://users.cecs.anu.edu.au/~bdm/data/graphs.html
23. GitHub code repository for the numerical experiments of the article. https://github.com/ilya160897/Random-walk-with-jumps

QAP Analysis of Company Co-mention Network

S. P. Sidorov, A. R. Faizliev[✉], V. A. Balash, A. A. Gudkov,
A. Z. Chekmareva, M. Levshunov, and S. V. Mironov

Saratov State University, Saratov, Russian Federation
faizlievar1983@mail.ru
http://www.sgu.ru

Abstract. In our research we form a network called company co-mention network. News analytics data have been employed to collect the companies co-mentioning. Each company acquires a certain value based on the amount of news in which the company was mentioned. A matrix containing the number of co-mentioning news between pairs of companies has been created for network analysis. Each company is presented as a node, and news mentioning two companies establishes a link between them. The network is constructed quite similarly to social networks or co-citation networks. The networked map of the companies is used to visualize the dependence structure of the economy by identifying groups of companies that are more central than others. The analysis carried out in the context of sectors of economy and territorial affiliation made it possible to identify key companies and to explore the similarity of the power law of vertices within sectors. QAP analysis between the co-mention network and the sector affiliation network was carried out to examine the ability of the sector affiliation network to predict the structure of the co-mention network.

Keywords: Network analysis · News analytics · Degree distribution
SNA metrics · QAP analysis

1 Introduction

One of the examples of big data that has emerged relatively recently is information flows generated by news agencies, enterprises, organizations, social networks, etc. Such real time information flows feature a huge amount of news sources, unstructurability, a large amount of both sources and objects of news, and high frequency (thousands of news items per second). Therefore, one of the problems of information systems that process such data is to aggregate them into one or several indicators that allow to describe the intensity, stability, changing news flow structure, identify the most discussed news subjects, and so on. Note that

This work was supported by the Russian Fund for Basic Research, project 18-37-00060.

the characteristics and features of the news flow are currently not sufficiently studied, and the methods and algorithms for processing such data are not fully developed.

In this paper we use some methods of the graph theory and corresponding algorithms to examine news flow data. A social network analysis (SNA) approach allows to investigate (explain) structures in systems based on the relations among the system's components (nodes) rather than the attributes of individual cases [6]. Methods of social network analysis can be used to analyze the structure of relations in an organization [7,8] or examine relationships among social entities [44]. The basic concepts of SNA are node and link, where a node refers to a unit (individual, object, item) and a link indicates a relationship between nodes.

News flow data can be easily converted into network data, since a company can be presented as a node and the act of mentioning two companies in one news item can be visualized as a link between them. Moreover, using this co-mention network we can study basic properties of networks, such as centrality and tie strength. Highly mentioned companies may be treated as key companies which are more significant to economy than other companies. Central nodes in this co-mention network are key companies in co-mention analysis. Moreover, we can treat the amount of co-mentions as the link weight. In this case, highly mentioned companies will have a higher value of link weight.

The paper [35] used various types of social network analysis metrics and citation indices to find key companies in the network. The focus of this paper is the analysis of different parts of co-mention network (or subnetworks) rather than the co-mention network as the whole. In our study, we use two ways to construct subnetworks:

– subnetworks represent different sectors of economy, such as products, resources, services, IT sectors.
– each subnetwork consists of companies with the same stock exchange affiliation.

The concept of link weight is crucial to our analysis. Through examination frequencies and other SNA metrics of each company within a subnet, co-mention analysis of companies can identify the key companies of the subnetwork. For co-mentioned companies, the weight of links may be useful for identification of the most frequently co-mentioned pairs of companies, as well as those who are poorly co-mentioned but useful in terms of providing diversity to the sector or subnetwork where they belong. We suppose that companies with weak links to other companies may play a valuable role in expanding the diversity of economic information within their subnetwork or clique.

In our research we would like to find answers to the following questions:

– Do the frequency, degree centrality, closeness centrality, betweenness centrality and eigenvector centrality of companies vary within subnetworks (clusters)? Within each group (or subnetwork) of companies, our research is going to find those which are more central than others. It is assumed that the more central a company is within a network, the more influential and more important the news about it must be.

- Does the type of the degree distribution vary within clusters? What type of the functional form has the clustering-degree relation for the clusters?
- Does the analysis of company co-mention network identify groups? Our hypothesis is that the network analysis of the company co-mention network reproduces the sector structure of the economy. The company co-mention network may be very sparse, but the graph of companies is expected to show which of them are often mentioned together or which belong to the same sector of economy. Each group or cluster of companies might be associated with a particular sector, for example, products, resources, services or information technology. Another hypothesis we would like to examine is that the structure of the company co-mention network reflects the clusterization of companies based on their stock exchange affiliation.

To provide answers to the first and the second questions we will use well-known SNA metrics and methods. The answer to the third question will rely on the quadratic assignment procedure (QAP) which was proposed and developed in [13,18,21,27]. Since then, QAP has been widely used in social network analysis (see e.g. [5,9,10,17,34,42], among many others). QAP is a peculiar type of permutation test which leaves the dyadic data structure under the permutations unchanged.

Our research questions are close to ones from the paper [20] which investigated whether the patterns of author co-citation can describe the structure of the field of communication.

Our study uses the data delivered by the news analytics providers. In our opinion, the data are quite typical and can be used for processing and analysis. Using this data, papers [36–39] examined different news flow characteristics, such as intensity, stability, volatility, long-term memory, fractality, etc.

2 Data

A huge amount of economic and financial news are generated in real time by news agencies, stocks exchanges, companies, magazines, papers, blogs and so on. Different companies are named in these news items. We accomplish company co-mention analysis in the five steps:

1. we assemble all economic and financial news items produced during one month of 2015 (February of 2015);
2. we collect a list of mentioned companies for each news item;
3. we calculate a weighted co-mention count for each pair of co-mentioned companies based on all set of available news items;
4. we produce a symmetric co-mention matrix using these weighted co-mention counts;
5. then we analyze this co-mention matrix statistically, and the results are visualized and interpreted.

Step 1 and 2 operations are executed by providers of news analytics. In our research, we manipulate these already processed data.

2.1   News Analytics Data

Two of the biggest providers of news analytics are Thompson Reuters and Raven Pack. They gather news items from diverse sources in real time. They collect data from different sources including news agencies and social media (blogs, social networks, etc.). They also use so-called pre-news, i.e. SEC reports, court documents, reports of various government agencies, business resources, company reports, announcements, industrial and macroeconomic statistics. Then news analytics providers handle preliminary analysis of each news item in real time. Using AI algorithms, they calculate news-related expectations (sentiments) based on the current market situation. As a rule, providers of news analytics provide to subscribers in real time the following attributes for each news item: time stamp, company name, company id, relevance of the news, event category, event sentiment, novelty of the news, novelty id, composite sentiment score of the news, among others. Subscribers of news analytics data may develop and exploit quantitative models or trading strategies based on both the news analytics data and financial time series data. The survey of applications for news analytics tools can be found in books [28, 29].

2.2   Generating Company Co-mention Network

Methodology. Company co-mention network is formed based on co-mention; it means that a company has connection with those companies that have been mentioned in a news item together. A company co-mention network is a set of companies which have connections in pair to represent their co-mention relationship. Two companies are linked if there has been published a publicly available news item mentioning both of them. In such type of network, a company will be represented by 'node' or 'vertex' and the connection will be an 'edge'. Thus, we represent the company co-mention network as an undirected weighted graph. In some sense, company co-mention network can be considered as social network. Based on the available data of news analytics, we built an adjacency matrix which represents the relationship between companies in line with the approach described in [35].

Network. We deal with all the financial and economic news items released during one month period from February 1, 2015 to February 28, 2015 (i.e. 20 trading days). We eliminated all the news on the imbalance of supply and demand before both the opening and the closing of trading time of different stock exchanges. News of such type may amount to several hundreds of news coming out in a short time at the beginning and at the end of the trading sessions. During February 2015, there were published more than 230 thousand news items which mentioned more than 18,000 companies. We assembled the list of all companies which have at least one common news item with at least one other company. There are more than 7,000 such companies during February, 2015.

Table 1 presents the descriptive statistics of time series.

Table 1. The descriptive statistics, one day

	All news	Stock exchanges			Sectors			
		NYSE	LSE	Tokyo SE	Products	Resources	Services	IT
Mean	8383.43	2351.57	902.43	730.75	1772.21	1348.50	1896.46	505.61
St. dev.	5415.21	1569.98	596.61	524.85	1159.12	928.81	1235.39	356.55
Skewness	−0.60	−0.49	−0.63	−0.22	−0.54	−0.17	−0.56	−0.31
Kurtosis	1.75	1.64	1.88	1.89	1.74	2.13	1.76	1.76

After obtaining non-directional symmetric matrix with valued weights for the co-mention counts of each pair of companies, we use R packages for finding basic statistics and to visualize the network of companies.

2.3 Social Network Analysis

Social Network Analysis (SNA) describes social relations in terms of graph theory. SNA presents objects (e.g. individuals, groups, organizations, URLs, and other connected entities) within the network as nodes, and links represent relationships (e.g. friendship, co-authorship, organizations and sexual relationships) between the objects [1, 14, 22, 25, 40].

Social networks can be represented in the form of a diagram, where nodes are points and links are lines. SNA deals with measuring relationships between objects [12, 26, 30, 44].

The nodes in our network represent companies and the links represent co-mention relationships between the nodes. SNA provides both visual and mathematical analysis of relationships [33, 41].

Recent years have seen increased interest in the study of Social Media using SNA (see e.g. [3, 11, 23, 45, 47], among many others).

Key objects are those that are in relationships with many other objects. In the context of our analysis, a company with extensive links or co-mention with many other companies in the economy is considered more important than a company with relatively fewer links. Different types of SNA metrics can be used to find key companies in the network. In our analysis, we use the following well-known metrics: degree centrality, closeness centrality, betweenness centrality, eigenvector centrality, frequency. A detailed description of these metrics can be found in the article [25].

2.4 Key Company Analysis According to Sectors of Economics

In addition to the co-mention matrix, we create a matrix describing which of ten economic sectors the companies belongs to (extraction of consumer discretionary, consumer staples, energy, financial, health care, industrials, information technology, raw materials, telecommunications services and utilities) and which stock exchange the companies is related to.

All the companies we considered were divided into four sectors of the economy in the following way:

$$\left.\begin{array}{r}\text{Consumer discretionary}\\\text{Consumer staples}\\\text{Industrials}\end{array}\right\}\longrightarrow \text{Products (2007 companies)}$$

$$\left.\begin{array}{r}\text{Energy}\\\text{Raw materials}\end{array}\right\}\longrightarrow \text{Resources (1398 companies)}$$

$$\left.\begin{array}{r}\text{Financials}\\\text{Health care}\\\text{Telecommunications}\\\bullet \quad \text{Utilities}\end{array}\right\}\longrightarrow \text{Services (2375 companies)}$$

$$\text{Information technology}\}\longrightarrow \text{IT sector (548 companies)}$$

Table 2 lists the top-5 companies with the highest frequency of co-mention in each of the 4 sectors. The table also shows the number of company's links and different standardized centrality indicators. As you can see from this table, General Motors with high frequency and with high degree is a key company in the products sector. It strongly dominates over the rest of the companies, so that it has the largest number of news co-mentions and links to other companies. We noted that in terms of proximity (closeness centrality), all considered companies are identical. This is typical for all sectors of the economy. Most of the largest companies in this sector belong to auto groups or aircraft manufactory.

Table 2 also shows the largest companies in the Resources sector. The leading companies in this sector are oil and gas companies, which is reasonable. At the same time American companies Apache Corporation and Continental Resources lead in co-mention frequency and number of links. The high of the Eigenvector centrality indicator proves that Apache Corporation and Continental Resources interact with other large companies in the sector. It acts as the bridge more often (it is a connecting link) connecting the companies of this sector.

In the services sector, there are several leaders which are the largest financial holdings: JPMorgan Chase, Citigroup, Bank of America. This group of companies are also leaders in all measures of centrality considered. Apple is the leader in the information technology sector for all of the indicators with a large margin. Earlier, in our article [35] it was shown that Apple is the leading (key) company among all the companies under consideration.

2.5   Key Company Analysis According to Stock Exchanges

Next, a network of co-mentions of companies belonging to different territorial zones was studied. We considered separately companies that are trading on the European, American and Asian stock exchanges. Precisely, we analyzed companies that traded on London SE (488 companies), New-York SE (1,715 companies) and Tokyo SE (586 companies).

Table 3 shows that the key companies of the London Stock Exchange in terms of the number of links are Barclays Bank PLC and Aviva PLC. Mining

Table 2. Companies with higher frequency for four sectors of the economy

Company	Frequency	Degree	Degree centrality $\times 10^1$	Closeness centrality $\times 10^3$	Betweenness centrality $\times 10^1$	Eigenvector centrality
Products						
General Motors Co	1051	143	0.71	2.0745	0.72	1.000
Ford Motor Co	691	90	0.45	2.0731	0.26	0.877
Volkswagen AG	668	101	0.50	2.0734	0.23	0.572
Boeing Co	663	103	0.51	2.0735	0.34	0.084
Bayerische Motoren Werke AG	652	108	0.54	2.0724	0.17	0.576
Rescources						
Apache Corp	1583	118	0.84	2.5374	0.04	0.966
Continental Resources Inc	1538	101	0.72	2.5373	0.06	1.000
Pioneer Natural Resources Co	1422	70	0.50	2.5354	0.00	0.984
RSP Permian Inc	1369	67	0.48	2.5353	0.00	0.937
Devon Energy Corp	1282	92	0.66	2.5356	0.01	0.849
Services						
JPMorgan Chase & Co	882	137	0.58	1.6921	0.43	1.000
Citigroup Inc	837	128	0.54	1.6922	0.43	0.889
Bank of America Corp	757	142	0.60	1.6926	0.69	0.763
Goldman Sachs Group Inc	706	105	0.44	1.6915	0.13	0.868
American Express Co	622	126	0.53	1.6921	0.39	0.391
IT sector						
Apple Inc	805	127	2.32	6.1457	2.59	1.000
Alphabet Inc	472	67	1.22	6.1364	0.79	0.927
Twitter Inc	371	54	0.99	6.1293	0.27	0.640
Samsung Electronics Co Ltd	336	56	1.02	6.1364	0.79	0.618
Facebook Inc	307	47	0.86	6.1323	0.32	0.640

company BHP Billiton PLC has the largest number of co-mentioned news and interacts with the largest companies of this exchange. Further in the table the key companies of the New York Stock Exchange are given; all of them belong to the oil and gas industry. At the same time key companies of the Tokyo Stock Exchange relate to engineering and industrial sectors.

2.6 Degree Distribution Analysis

In this section, the distribution analysis for the sub-graphs on 4 sectors of the economy and for the 3 stock exchanges is provided. The description of this procedure can be found in the articles [2, 19].

Table 3. Companies with higher frequency for three stock exchanges

Company	Frequency	Degree	Degree centrality $\times 10^1$	Closeness centrality $\times 10^5$	Betweenness centrality $\times 10^2$	Eigenvector centrality
London SE						
Barclays PLC	484	42	1.72	3.2481	3.37	0.374
Aviva PLC	429	47	1.93	3.2517	5.20	0.308
BHP Billiton PLC	363	81	3.33	3.2576	4.29	1.000
BP PLC	313	53	2.18	3.2511	3.43	0.536
Associated British Foods PLC	303	68	2.79	3.2534	3.00	0.733
New-York SE						
Continental Resources Inc	1594	178	2.08	0.2767	0.50	0.720
Apache Corp	1588	184	2.15	0.2774	1.02	0.733
Anadarko Petroleum Corp	1386	217	2.53	0.2780	1.96	0.831
Basic Energy Services Inc	1336	252	2.94	0.2778	1.05	1.000
Halliburton Co	1281	247	2.88	0.2778	1.52	0.924
Tokyo SE						
Bombardier Inc	515	154	5.26	2.0368	6.32	0.994
Calfrac Well Services Ltd	489	105	3.59	2.0299	0.59	0.710
Alara Resources Ltd	464	120	4.10	2.0320	1.08	0.883
Newalta Corp	428	147	5.03	2.0345	1.86	1.000
Strad Energy Services Ltd	414	101	3.45	2.0305	0.68	0.633

A degree distribution is called power-law distribution if

$$n(k) \sim Ak^{-\gamma},$$

where γ is degree exponent.

For all stock exchanges under review (Table 5) and economic sectors (Table 4) the distribution of degrees follows the power law. Figure 1 shows the dependence of the number of companies $n(k)$ on node degree k for NYSE.

The resulting models are statistically significant at alpha level of 0.01. However, the degree exponent for all stock exchanges and sectors of the economy does not fall in the interval $(2, 3)$. Thus, the decrease in the degrees of vertices is slower than for typical social networks [16,24]. It is also notable that the fewer companies are in a subgraph, the slower the degree of vertices decreases are and the less the coefficient of determination is.

Table 4. Degree distribution for four sectors of the economy

Subgraphs	Degree exponent γ	R^2
Products	1.06	0.82
Rescources	0.91	0.79
Services	1.14	0.84
Information technology	0.64	0.65

Table 5. Degree distribution for three stock exchanges

Subgraphs	Degree exponent γ	R^2
London SE	1.10	0.81
New-York SE	1.18	0.84
Tokyo SE	0.68	0.44

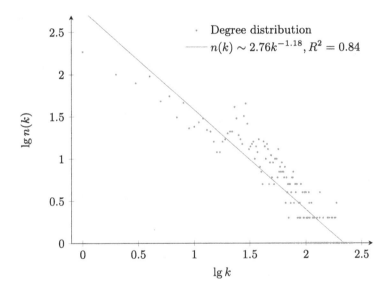

Fig. 1. The degree distribution of the New-York's companies co-mention network

2.7 Clustering Coefficient Distribution

In this section, clustering coefficient distributions analysis will be carried out for sub-graphs in 4 sectors economy and for 3 stock exchanges described above. Description of these procedures can be found in articles [4, 15, 43].

The average clustering coefficient of nodes $C(k)$ with degree k has been found:

$$C(k) \sim Bk^{-\beta},$$

where the exponent β usually lies between 1 and 2 [31, 32, 46].

For the given networks, the clustering-degree distribution relation follows the power law.

The resulting models are statistically significant at alpha level of 0.01. Herewith, the exponent β is turned out less than 1 for all the subgraphs under consideration (Tables 4 and 5).

It should also be noted that the power dependence between the local clustering coefficient and the degree is manifested for sufficiently large degrees of the vertices of k. There is no dependence for relatively small k. This fact is typical for all subgraphs under consideration and it can be well observed in the example of a subgraph (Fig. 2). Herewith, the exponent β is turned out less than 1 for all stock exchanges and sectors of the economy (Tables 6 and 7).

Sometimes the flow contains news which mention a large number of actors (companies). For example, about 0.5% of news reports allude to 10 or more companies. In such cases, the procedure we used to fill the incidence matrix generated a "pleiad" - a subgraph that included all possible verges between the actors mentioned in the report, which led to deviations of the vertex degree and clustering coefficient values from the general pattern. Note that exclusion of a large number of co-occurrences from the news analysis eliminated this problem. However, this method leads to a significant loss of information. It seems more accurate to use a modified procedure for clustering coefficient calculation that takes into account the peculiarities of the co-occurrences flow. We are planning to do this in the future.

In further research, we are taking into account this feature of the news flow, particularly in procedures for the news flow filtering and decomposing. We propose to allocate and consider repetitive co-mentions (a stable part of the graph in time), co-mentions caused by certain events in politics or economy (eventual co-mentions), and, finally, random perturbations separately.

Table 6. Local clustering coefficient for four sectors of the economy

Subgraphs	Exponent β	R^2
Products	0.70	0.58
Rescources	0.44	0.46
Services	0.70	0.61
Information technology	0.61	0.54

Table 7. Local clustering coefficient for three stock exchanges

Subgraphs	Exponent β	R^2
London SE	0.68	0.55
New-York SE	0.68	0.74
Tokyo SE	0.78	0.74

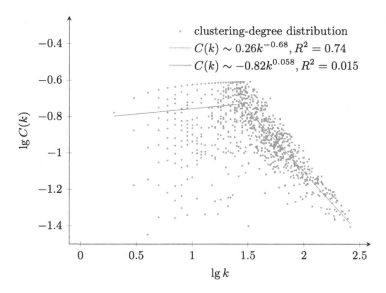

Fig. 2. The clustering-degree distribution of the New-York's companies co-mention network

3   QAP Correlation and Regression Analysis

Using the co-mention matrix and the companies' sector affiliation matrix (as well as the stock exchange affiliation matrix) we conduct QAP Correlation Analysis. QAP (Quadratic Assignment Procedure) was proposed and developed in [13,18, 21,27]. We use QAP Correlation Analysis to identify correlations

- between the co-mention network and companies' sector affiliation,
- between the co-mention network and stock exchange affiliation.

With the co-mention network as a prime network, corresponding cells of the sector affiliation matrix (as well as the stock exchange affiliation matrix) are compared to compute the value of Pearson's correlation. We repeat the process randomly permuting columns and rows to find the correlation. A lower value of Pearson's correlation means a stronger relationship between the matrices.

The first research hypothesis states that the graph of companies resulting from a network analysis identify sectors. First, QAP analyses between the co-mention network and the sector affiliation network was carried out to examine whether the sector affiliation network can predict the structure of the co-mention network. We calculate the value of Pearson correlation using R package.

The QAP correlation analysis shows a significant correlation between the co-mention network and the stock exchange affiliation ($r = 0.053$, $p = 0.000$) and the sector affiliation network ($r = 0.020$, $p = 0.000$).

We exploited the QAP procedure for testing the significance of the correlation coefficients. Estimated density of QAP replications for the sector affiliation network is shown in Fig. 3. Similar results were also obtained for the network of co-mentions and the Stock exchange affiliation. The observed values of the correlation coefficients were higher than the model values in all simulated 500 samples. Thus, the observed correlation coefficients are statistically significant, while they are close to zero. This can be explained by the fact that the adjacency matrices were of large dimension and were sufficiently discharged. This could contribute to the underestimation of the correlation coefficient.

We evaluated the linear regression between the elements of the co-mention network matrix and the stock exchange affiliation matrix, as well as elements of the co-mention network matrix and the sector affiliation matrix. The values of the parameters found by the least squares method and p-value obtained by the QAP procedure are given in Table 8.

The QAP regression analysis shows (Table 8), that only 0.2% variance is predicted by the model ($R^2 = 0.002$). This value is relatively low, and indicates an insufficient inclusion of explanatory variables. The coefficients of the model are statistically significant.

Table 8. QAP regression analysis

Coefficients	Estimate	P-value
Intercept	0.003	0.000
Sector affiliation	0.015	0.000
Stock exchange affiliation	0.033	0.000

The co-mention network of companies has similar structure to their territorial connections. We visualize the co-mention matrix using R package. We identify clusters using the level of link weights which is derived from co-mention frequencies. We use the sector affiliation matrix as attribute data to see if companies of one cluster have the attribute in common. Figure 4 shows a small part of network map of companies. Nasdaq companies with New York Stock Exchange and Dax companies with London Stock Exchange make two clusters. London companies along with GM (General Motors) are more likely to be a bridge for the US connection with Europe.

Big companies generate a much bigger news flow than small companies. For this reason, the co-mention matrix is dense for the largest companies and is much sparser for small companies. However, the overall spatial and sector affiliation of companies significantly affects the probability and frequency of co-mentions. In our opinion, the widening of the range of analyzed companies induces the sparsity of the co-mention matrix, which leads to a drop in the share of the explained variance.

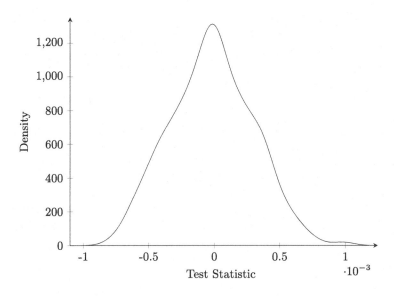

Fig. 3. Estimated density of QAP replications for the sector affiliation network

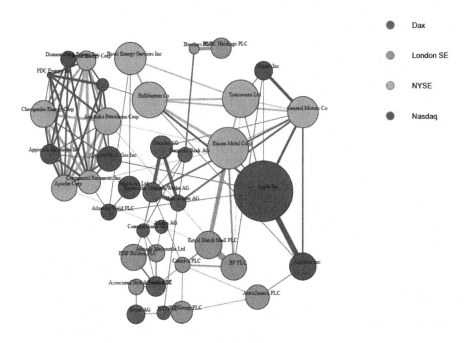

Fig. 4. Networked Map of companies

4   Conclusion

In this article, we investigated the relationship between company co-mention network and the sector affiliation matrix. Moreover, we identified key companies in different sectors of the economy using various indicators of network analysis, such as frequency, normalized degree of centrality, closeness centrality, betweenness centrality and eigenvector centrality. We discovered that different network analysis indicators show different values for different companies. But some of the companies have high significance for all indicators considered. At the same time, the majority of leading (key) companies belong to the New York Stock exchange. It was shown that the distribution of degrees and clustering-degree relations for our network adheres to the power law, although with nonstandard indicators of exponent. QAP analysis showed the presence of significant positive correlation between company co-mention network and stock exchange affiliation, and between company co-mention network and sector affiliation network.

References

1. Abbasi, A., Altmann, J.: On the correlation between research performance and social network analysis measures applied to research collaboration networks. In: 44th Hawaii International Conference on System Sciences (HICSS), pp. 1–10. IEEE (2011)
2. Albert, R., Barabasi, A.L.: Statistical mechanics of complex networks. Rev. Mod. Phys. **74**, 47–97 (2002)
3. An, J., Kwak, H.: What gets media attention and how media attention evolves over time: large-scale empirical evidence from 196 countries. In: Proceedings of the Eleventh International Conference on Web and Social Media, pp. 464–467. The AAAI Press, Palo Alto, Montreal, May 2017
4. Anthonisse, J.M.: The rush in a directed graph. Technical (1971)
5. Atzmueller, M., Schmidt, A., Kloepper, B., Arnu, D.: HYPGRAPHS: an approach for analysis and assessment of graph-based and sequential hypotheses. In: Appice, A., Ceci, M., Loglisci, C., Masciari, E., Raś, Z.W. (eds.) NFMCP 2016. LNCS (LNAI), vol. 10312, pp. 231–247. Springer, Cham (2017). https://doi.org/10.1007/978-3-319-61461-8_15
6. Barnett, G.: A longitudinal analysis of the international telecommunication network, 1978–1996. Am. Behav. Sci. **44**, 1638–1655 (2001)
7. Barnett, G., Danowski, J.: The structure of communication: a network analysis of the international communication association. Hum. Commun. Res. **19**(2), 264–285 (1992)
8. Barnett, G., Salisbury, J.: Communication and globalization: a longitudinal analysis of the international telecommunication network. J. World Syst. Res. **2**(16), 1–17 (1996)
9. Basov, N., Lee, J.S., Antoniuk, A.: Social networks and construction of culture: a socio-semantic analysis of art groups. In: Cherifi, H., Gaito, S., Quattrociocchi, W., Sala, A. (eds.) Complex Networks & Their Applications V. COMPLEX NETWORKS 2016, vol. 693, pp. 785–796. Springer, Cham (2017). https://doi.org/10.1007/978-3-319-50901-3_62

10. Choi, E., Lee, K.C.: Relationship between social network structure dynamics and innovation: micro-level analyses of virtual cross-functional teams in a multinational B2B firm. Comput. Hum. Behav. **65**, 151–162 (2016)
11. Coletto, M., Garimella, K., Gionis, A., Lucchese, C.: A motif-based approach for identifying controversy. In: Proceedings of the Eleventh International Conference on Web and Social Media, pp. 496–499. The AAAI Press, Palo Alto, Montreal, May 2017
12. Correa, C., Crnovrsanin, T., Ma, K.L.: Visual reasoning about social networks using centrality sensitivity. IEEE Trans. Vis. Comput. Graph. **18**(1), 106–120 (2012)
13. Dekker, D., Krackhardt, D., Snijders, T.A.B.: Sensitivity of MRQAP tests to collinearity and autocorrelation conditions. Psychometrika **72**(4), 563–581 (2007)
14. Deville, P., Song, C., Eagle, N., Blondel, V.D., Barabási, A.L., Wang, D.: Scaling identity connects human mobility and social interactions. Proc. Nat. Acad. Sci. **113**(26), 7047–7052 (2016). http://www.pnas.org/content/113/26/7047
15. Granovetter, M.: The strength of weak ties. Am. J. Sociol. **78**, 1360 (1973)
16. Guelzim, N., Bottani, S., Bourgine, P., Kepes, F.: Topological and causal structure of the yeast transcriptional network. Nat. Genet. **31**, 60–63 (2002)
17. Haishu, Q., Ying, L., Xin, O.: Industrial association, common information spill out and industry stock indexes co-movement. Syst. Eng. Theory Pract. **36**(11), 2737 (2016)
18. Hubert, L.: Assignment Methods in Combinatorial Data Analysis. Dekker, New York (1987)
19. Jeong, H., Tombor, B., Albert, R., Oltvai, Z.N., Barabasi, A.L.: The large-scale organization of metabolic networks. Nature **407**, 651–654 (2000)
20. Kim, H., Barnett, G.A.: Social network analysis using author co-citation data. In: AMCIS 2008 Proceedings, Paper 172, pp. 1–9 (2008)
21. Krackardt, D.: Qap partialling as a test of spuriousness. Soc. Netw. **9**(2), 171–186 (1987)
22. Landherr, A., Friedl, B., Heidemann, J.: A critical review of centrality measures in social networks. Bus. Inf. Syst. Eng. **2**(6), 371–385 (2010)
23. Le, H., Shafiq, Z., Srinivasan, P.: Scalable news slant measurement using twitter. In: Proceedings of the Eleventh International Conference on Web and Social Media, pp. 584–587. The AAAI Press, Palo Alto, Montreal, May 2017
24. Lee, T.I., Rinaldi, N.J., Robert, F., Odom, D.T., Bar-Joseph, Z., Gerber, G.K., Hannett, N.M., Harbison, C.T., Thompson, C.M., Simon, I., Zeitlinger, J., Jennings, E.G., Murray, H.L., Gordon, D.B., Ren, B., Wyrick, J.J., Tagne, J.B., Volkert, T.L., Fraenkel, E., Gifford, D.K., Young, R.A.: Transcriptional regulatory networks in saccharomyces cerevisiae. Science **298**(5594), 799–804 (2002)
25. Liu, B.: Web Data Mining. Springer, Heidelberg (2007). https://doi.org/10.1007/978-3-540-37882-2
26. Liu, X., Bollen, J., Nelson, M.L., de Sompel, V.: Co-authorship networks in the digital library research community. Inf. Process. Manag. **41**(6), 1462–1480 (2005)
27. Mantel, N.: The detection of disease clustering and a generalized regression approach. Cancer Res. **27**(2), 209–220 (1967)
28. Mitra, G., Mitra, L. (eds.): The Handbook of News Analytics in Finance. Wiley, Hoboken (2011)
29. Mitra, G., Yu, X. (eds.): Handbook of Sentiment Analysis in Finance (2016)
30. Onnela, J.P., Arbesman, S., Gonzalez, M.C., Barabasi, A.L., Christakis, N.A.: Geographic constraints on social network groups. PLoS ONE **6**(4), 1–7 (2011). https://doi.org/10.1371/journal.pone.0016939

31. Ravasz, R., Barabasi, A.L.: Hierarchical organization in complex networks. Phys. Rev. E **67**, 026112 (2003)
32. Ravasz, R., Somera, A.L., Mongru, D.A., Oltvai, Z.N., Barabasi, A.L.: Hierarchical organization of modularity in metabolic networks. Science **297**, 1551–1555 (2002)
33. Said, Y.H., Wegman, E., Sharabati, W.K., Rigsby, J.: Social networks of author-coauthor relationships. Comput. Stat. Data Anal. **52**(4), 2177–2184 (2008)
34. Samoilenko, A., Karimi, F., Edler, D., Kunegis, J., Strohmaier, M.: Linguistic neighbourhoods: explaining cultural borders on wikipedia through multilingual co-editing activity. EPJ Data Sci. **5**, 1–20 (2016)
35. Sidorov, S.P., Faizliev, A.R., Balash, V.A., Gudkov, A.A., Chekmareva, A.Z., Anikin, P.K.: Company co-mention network analysis. Springer Proceedings in Mathematics and Statistics (2018, in press)
36. Sidorov, S., Faizliev, A., Balash, V.: Measuring long-range correlations in news flow intensity time series. Int. J. Mod. Phys. C **28**(08), 1750103 (2017)
37. Sidorov, S., Faizliev, A., Balash, V.: Scale invariance of news flow intensity time series. Nonlinear Phenom. Complex Syst. **19**(4), 368–377 (2016)
38. Sidorov, S., Faizliev, A., Balash, V.: Fractality and multifractality analysis of news sentiments time series. IAENG Int. J. Appl. Math. **48**(1), 90–97 (2018)
39. Sidorov, S., Faizliev, A., Balash, V., Korobov, E.: Long-range correlation analysis of economic news flow intensity. Phys. A **444**, 205–212 (2016)
40. Sinatra, R., Wang, D., Deville, P., Song, C., Barabási, A.L.: Quantifying the evolution of individual scientific impact. Science **354**(6312), aaf5239 (2016). http://science.sciencemag.org/content/354/6312/aaf5239
41. Tang, J., Zhang, D., Yao, L.: Social network extraction of academic researchers. In: Seventh IEEE International Conference on Data Mining, pp. 292–301. IEEE (2007)
42. Vahtera, P., Buckley, P.J., Aliyev, M., Clegg, J., Cross, A.R.: Influence of social identity on negative perceptions in global virtual teams. J. Int. Manag. **23**(4), 367–381 (2017)
43. Wagner, A., Fell, D.A.: The small world inside large metabolic networks. Proc. R. Soc. Lond. B Biol. Sci. **268**, 1803–1810 (2001)
44. Wasserman, S., Faust, K.: Social Network Analysis: Methods and Applications. Cambridge University Press, Cambridge (1994)
45. West, R., Pfeffer, J.: Armed conflicts in online news: a multilingual study. In: Proceedings of the Eleventh International Conference on Web and Social Media, pp. 309–318. The AAAI Press, Palo Alto, Montreal, May 2017
46. Yook, S.H., Oltvai, Z.N., Barabasi, A.L.: Functional and topological characterization of protein interaction networks. Proteomics **4**, 928–942 (2004)
47. Zhang, A., Culbertson, B., Paritosh, P.: Characterizing online communities using coarse discourse structures. In: Proceedings of the Eleventh International Conference on Web and Social Media, pp. 357–366. The AAAI Press, Palo Alto, Montreal, May 2017

Towards a Systematic Evaluation
of Generative Network Models

Thomas Bläsius, Tobias Friedrich, Maximilian Katzmann$^{(\boxtimes)}$, Anton Krohmer,
and Jonathan Striebel

Hasso Plattner Institute, Potsdam, Germany
`maximilian.katzmann@hpi.de`

Abstract. Generative graph models play an important role in network
science. Unlike real-world networks, they are accessible for mathematical analysis and the number of available networks is not limited. The
explanatory power of results on generative models, however, heavily
depends on how realistic they are. We present a framework that allows
for a systematic evaluation of generative network models. It is based on
the question whether real-world networks can be distinguished from generated graphs with respect to certain graph parameters.

As a proof of concept, we apply our framework to four popular random
graph models (Erdős-Rényi, Barabási-Albert, Chung-Lu, and hyperbolic
random graphs). Our experiments for example show that all four models
are bad representations for Facebook's social networks, while Chung-Lu
and hyperbolic random graphs are good representations for other networks, with different strengths and weaknesses.

Keywords: Generative graph models · Real-world comparison
Distinguishability of network classes

1 Introduction

Generative graph models play an important role in network science for a multi-
tude of reasons. They can explain how certain properties observed in real-world
networks naturally emerge when assuming simplified but reasonable creation
mechanisms. The *small-world phenomenon* for example emerges from a small
amount of randomness in the form of independently chosen edges [10,23], and
analyzing how information spreads in random networks can help to explain *information cascades*, in which individuals act based on the behavior of other individuals instead of their own information, leading to a herd-like behavior [22].
Moreover, analyzing the expected run time of an algorithm on a realistic generative model has the potential to explain why certain algorithms perform well
on real-world instances despite their bad worst-case performance [16]. Finally,
randomly generated instances can serve as benchmark sets for algorithms.

The explanatory power of a generative model and its usefulness as benchmark
heavily depends on how well the generated graphs mimic real-world networks.

© Springer International Publishing AG, part of Springer Nature 2018
A. Bonato et al. (Eds.): WAW 2018, LNCS 10836, pp. 99–114, 2018.
https://doi.org/10.1007/978-3-319-92871-5_8

A common way to provide evidence for the usefulness of a model, is to analyze it with respect to certain fundamental properties. The properties commonly perceived as most important are the *degree distribution* (which is typically heterogeneous with many vertices of low degree and few vertices of high degree), the *diameter* (maximum distance between nodes, which is typically small), and the *clustering coefficient* (providing a measure of locality, which is typically high).

Typical examples of arguments for or against a certain model are as follows. The Barabási-Albert model leads to a power-law degree distribution, which is realistic for certain classes of real-world networks [3,7]. Chung-Lu graphs have the small-world property often observed in real-world networks as their diameter is rather small (namely $\Theta(\log n)$) [8]. The clustering coefficient of Barabási-Albert graphs tends to 0 for $n \to \infty$ [11], while it is bounded away from 0 for hyperbolic random graphs [15,17], making the latter more realistic.

Though knowing the asymptotic behavior of these fundamental properties is an important contribution to understand a model, there are disadvantages when it comes to judging how realistic it is: the statements are only of qualitative nature (a parameter is "small" or "large") but it is unclear which values are actually realistic. This is particularly true when trying to compare the asymptotic growth in a model with the specific numbers of a few real-world networks.

A more direct comparison is achieved by comparing how different a generated network and its real-world counterpart are. Such an approach heavily depends on the used similarity measure [20]. While these measures have important applications, they typically compare only pairs of networks. Thus, we believe they are not suited to evaluate the usefulness of a generative model as this pairwise comparison heavily favors overfitting and discourages the models to generalize.

The goal of being as unbiased as possible while mimicking certain important properties of real-world networks is formally captured by the term *maximum entropy model*. Erdős-Rényi graphs are for example maximum entropy with respect to the number of vertices and edges, i.e., each graph with the desired number of vertices and edges is produced with the same probability. Using this perspective, the perfect model would be one that is maximum entropy with respect to as few properties as possible such that the generated networks are indistinguishable from real-world networks with respect to as many properties as possible.

Our goal with this paper is to develop a framework that enables a systematic experimental evaluation of how good generative graph models are. In particular, it is possible to answer questions of the following type.

- Barabási-Albert, Chung-Lu, and hyperbolic random graphs all have small diameter ($\Theta(\log n / \log \log n)$ [6], $\Theta(\log n)$ [8], and polylogarithmic [14]). Which of the three is more realistic?
- Which aspects of real-world networks are well represented by a given model and which are not? Note that answering this question is particularly interesting for maximum entropy models, as it provides a direction on how to make the model more realistic.

- Which types of real-world networks (e.g., social or infrastructural) are well represented by a given model?
- Given two seemingly similar models, do they actually generate graphs with similar properties?

Contribution and Outline. We developed a framework capable of answering these questions. The general approach is to select generative models, a collection of real-world networks, and a set of parameters. For each model, a set of graphs fitted to the real-world networks is generated. We then answer the question whether the chosen set of parameters is sufficient to distinguish between the different collections using machine learning. This general question allows us to formulate all the above mentioned specific questions by appropriately choosing the graph collections that should be distinguished and the set of parameters.

The different components of the framework can be easily adapted: New real-world networks, further generative models, and additional parameters can be included. Moreover, the used machine learning technique is interchangeable.

To showcase our framework, we selected four models (Erdős-Rényi, Barabási-Albert, Chung-Lu, and hyperbolic random graphs) and evaluated them on 219 real-world networks based on ten different parameters. Our findings, interesting in their own right, are as follows.

- While all four models are bad representations for Facebook graphs, Chung-Lu and hyperbolic random graphs are reasonable models for other real-world networks.
- While the Chung-Lu model is better for features related to node degrees, hyperbolic random graphs excel when involving clustering or distance-related features.
- Though hyperbolic random graphs have a realistic average clustering, the variance in clustering is too low.
- In the Barabási-Albert model, the choice of the initial graph (clique or cycle) is only irrelevant if the average degree is small.

Our framework and the raw data produced in our experiments are available at https://github.com/jstriebel/nemo-eva.

Related Work. Attar and Aliakbary recently followed a similar approach of classifying networks based on certain graph parameters [1]. Their perspective is, however, significantly different: their goal is to decide for a given real-world network, which model is most suited to represent it. We note that this approach can also be used to evaluate which model is the most realistic for certain real-world networks by counting how many real-world networks are classified as which model. However, this only leads to a evaluation in comparison to the other models under consideration and it does not identify parameters with respect to which a model requires improvement.

2   Methodology

Our framework consists of three main steps. First, multiple collections of graphs are determined. Typically, we have one collection containing real-world networks and one collection for each generative model. In the second step, different graph parameters are computed for all graphs in a collection. For each graph, this yields a feature vector, which is used as its representation. Thus, the second step turns the collections of graphs into collections of feature vectors. The third step then determines, whether two collections can be distinguished based on the feature vectors, and if yes, which subsets of features can or cannot be used to distinguish between them. In the following, we describe the three steps in more detail.

2.1   Collections of Networks

In the first step, a collection of real-world networks and a selection of generative models is chosen. We denote the collection of real-world networks by $C = \{G_1, \ldots, G_c\}$ with $c = |C|$. For each model M and each graph $G_i \in C$, we use M to generate an artificial graph G_i^{M} trying to mimic the real-world network G_i. We denote the set of resulting networks by $C^{\mathrm{M}} = \{G_1^{\mathrm{M}}, \ldots, G_c^{\mathrm{M}}\}$.

Fitting the Models. We want the graph G_i^{M} generated by M to mimic the corresponding real-world network G_i. As this highly depends on the chosen model, it is not part of the framework. Most models, however, generate graphs based on a small set of input parameters and produce graphs that roughly match these parameters. In this case, we compute the relevant parameters for G_i and generate G_i^{M} using the resulting values. For input parameters of a model that are known to mainly influence one parameter of the generated graphs, without knowing an exact formula for this dependency, one can use a binary search to fit this parameter. In Sect. 3.2 we describe the fitting we used for Erdős-Rényi Graphs, Barabási-Albert, Chung-Lu, and hyperbolic random graphs.

2.2   Network Parameters

In the second step, each graph G is turned into a feature vector by computing the values of different parameters. Formally, a *feature* φ is a function that maps G to a numerical value $\varphi(G)$. For a feature set $F = \{\varphi_1, \ldots, \varphi_f\}$ of $f = |F|$ features, the *feature vector* of G is $(\varphi_1(G), \ldots, \varphi_f(G))$ and we denote it by $F(G)$. For a collection C of graphs, $F(C)$ denotes the corresponding collection of feature vectors. We note that selecting a sufficiently expressive set of parameters is crucial: our framework is based on the assumption that structural properties distinguishing different networks types can be represented by the chosen features.

Feature Cleaning. To eliminate meaningless features, we apply three data cleaning techniques: numerical cleaning, variation cleaning, and correlation grouping. The *numerical cleaning* eliminates all features that are undefined or infinite for at least one of the networks. The *variation cleaning* eliminates features that have little predictive value as they assume similar values on most networks. More precisely, features are eliminated based on their *normalized coefficient of variation*, which is defined as follows. For a given feature, let X be the vector containing the c values it assumes in different graphs. Then the feature's normalized coefficient of variation is defined as

$$\frac{\sigma(X)}{\mu(X)\sqrt{c-1}},$$

where σ and μ denote the standard deviation and the arithmetic mean, respectively. We remove features with a normalized coefficient of variation below a threshold of 1%.

The *correlation grouping* groups highly correlated features, as having multiple very similar features does not add any predictive value. For each group of correlated features only the feature with the clearest semantics (given by a manually predefined order) of the group is used. The grouping is done by constructing a graph, using the features as nodes and connecting two features by an edge if they have an absolute Spearman's rank correlation coefficient above 99%. Each connected component in that graph is one group. Note that grouped features can have a smaller correlation than the threshold of 99%, as the correlation is not transitive, but being in the same connected component is.

2.3 Distinguishing the Collections

In the third step, we want to determine which pairs of collections can be distinguished based on which features. To this end, we want to answer queries of the following type. The input is a subset F of all features and two collections of graphs, typically the collection C of real-world networks and the collection C^{M} for one model M (it is also possible to compare the collections of two different models, but for the sake of readability, we assume C and C^{M} in the following). We then want to know how well $F(C)$ can be distinguished from $F(C^{\text{M}})$, i.e., whether it can be learned which feature vectors are members of which collections by observing the membership only for few samples.

Classification Task. The input for the classifier consists of a feature-matrix $X \in \mathbb{R}^{2c \times f}$ ($c = |C|$ and $f = |F|$) and a binary vector $Y \in \{0,1\}^{2c}$ that classifies the features as belonging to C (denoted as 0) or as to C^{M} (denoted as 1). They are defined as

$$X = \begin{pmatrix} F(G_1) \\ \vdots \\ F(G_c) \\ F(G_1^{\text{M}}) \\ \vdots \\ F(G_c^{\text{M}}) \end{pmatrix} \quad \text{and} \quad Y = \begin{pmatrix} \left.\begin{matrix} 0 \\ \vdots \\ 0 \end{matrix}\right\} c \\ \left.\begin{matrix} 1 \\ \vdots \\ 1 \end{matrix}\right\} c \end{pmatrix}.$$

The task of distinguishing features X according to the vector Y corresponds directly to the classical machine-learning setting of binary classifications: Given the feature-space $\mathcal{X} = \mathbb{R}^{2c \times f}$ for $2c$ observations of c graphs and c corresponding models with f real-valued features, the target value space is defined as $\mathcal{Y} = \{0,1\}^{2c}$, and the binary classification model $M_\mathcal{X}$ is a function of the form $M_\mathcal{X} \colon \mathcal{X} \to \mathcal{Y}$. The output for each prediction is independent from other predictions, therefore

$$M_{\mathbb{R}^{2c \times f}}((x_1 \cdots x_{2c})^T) = (M_{\mathbb{R}^f}(x_1) \cdots M_{\mathbb{R}^f}(x_{2c}))^T.$$

Evaluation. To evaluate the resulting predictions, the accuracy is measured. Given the actual target values Y and the predicted values $\hat{Y} = M(X)$, the accuracy is the ratio of correctly classified examples [2]. Therefore,

$$\mathrm{accuracy}(\hat{Y}, Y) = \frac{\sum_i [\hat{Y}_i = Y_i]}{|Y|},$$

where $[\cdot]$ is the Iverson bracket with $[p] = 1$ if p is true and $[p] = 0$ otherwise.

Supervision and Cross-Validation. To do supervised learning, we need training data X_{train} and Y_{train}. With this, a supervised learning strategy S results in a trained model $M_\mathcal{X}$, therefore $S \colon \mathcal{X}_{\mathrm{train}} \times \mathcal{Y}_{\mathrm{train}} \to (M_\mathcal{X} \colon \mathcal{X} \to \mathcal{Y})$.

To make use of all the data as a target to predict, but simultaneously prevent to use the same data as a training and a testing example, we use cross-validation. Given some predictors X and target values Y, the ℓ-*fold cross-validation* splits the data in ℓ random, equally-sized subsets $X_1, \ldots, X_\ell, Y_1, \ldots, Y_\ell$. They are used to generate ℓ learned models, where for each model a single subset is used as the test dataset and all other subsets as the training data. To make the training unbiased we use *stratified cross-validation* which ensures that the number of examples is the same for both classes (i.e., each X_i includes the same number of feature vectors from $F(C)$ as from $F(C^M)$). The total accuracy of the cross-validation is then defined as the arithmetic mean of the accuracies of all models.

Classification Model. From the wide range of possible supervised machine learning classifiers we use support vector machines (SVMs) with the Gaussian radial basis function (rbf) kernel because they have a good predictive performance in general [13], are able to capture high order dependencies [4,19], and the parametrized regularization allows to tune the variance-bias trade-off [2,19].

To select the best parameters for the SVM and the rbf kernel, cross-validation over a grid of parameters is performed. The model with the best average accuracy in the testing sets is used as the final model. All features used in the SVM are normalized to have an arithmetic mean of zero and unit variance in the training data. The testing dataset is scaled using the same parameters. This ensures that also the scaling is done in an unbiased, cross-validated fashion.

3   Experiments

For the experiments we used 219 publicly available graphs from Network Repository [18]. For disconnected graphs, we used the largest connected component.

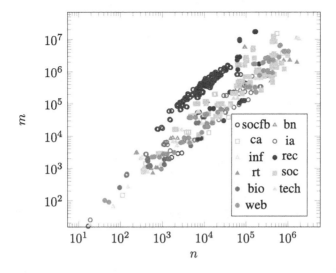

Fig. 1. The graphs used in our experiments.

The Network Repository divides graphs into different categories, as shown in Fig. 1. The graphs vary in their sizes and have up to 1.8 million nodes and 17 million edges. In the following, we define the used graph parameters, describe the considered generative models, and how we fitted them to the real-world networks.

3.1   Graph Properties

Table 1 lists all features we use. The properties we consider in our experiments can be divided into two categories: *single-value features* and *distributions over nodes*. We used NetworKit [21] to compute the features.

Single-Value Features. These are features that assign a single numerical value to a given graph. The most basic properties of this type are the *number of nodes* and the *number of edges*. Additionally, the *diameter* describes the maximum length of a shortest path between any two nodes in the graph and the *effective diameter* (a similar but more robust measure) represents an upper bound on the shortest path between 90% of all node pairs.

The generative models we consider are mostly meant to represent so-called scale-free networks whose degree distribution follows a power law, i.e., the fraction of vertices with at least k neighbors roughly behaves like $k^{-\beta}$, where β is the so-called *power-law exponent*.

Distribution over Nodes. These are features that assign a value to each node, leading to distributions over all nodes in the graph. For each of these distributions we consider the *arithmetic mean*, *median*, *first quartile* and *third quartile*, as well as the *standard deviation*.

The simplest measure of this type is the degree distribution, assigning each vertex its *degree*. The *local clustering coefficient* of a vertex v is the probability that two randomly selected neighbors of v are connected. The arithmetic mean of the local clustering coefficients is often referred to as "average local clustering coefficient" or simply "clustering coefficient" of the network and represents an important single-value feature. The k-core of a graph is obtained by successively removing all nodes with degree less than k. This leads to the measure of *core centrality*, where each node is assigned the largest k such that it is contained in the k-core. The *betweenness centrality* measures for each vertex v how many shortest paths between pairs of other nodes go though v, and the *closeness centrality* of a node denotes its average distance to every other node in the graph. Furthermore, the *Katz centrality* measures the importance of a node by its number of neighbors and the distance of all other nodes to these neighbors. Finally, the *PageRank centrality* is basically the limiting probability distribution of a random walk.

Table 1. The parameters we use, their abbreviations, and whether they are single-value or distribution parameters.

Feature	Abbreviation	SV/Distr.
Number of nodes	n	Single value
Number of edges	m	Single value
Diameter	d	Single value
Effective diameter	d'	Single value
Power-law exponent	β	Single value
Degree	deg	Distribution
Local clustering coefficient	c	Distribution
Core centrality	core	Distribution
Betweenness centrality	betw	Distribution
Closeness centrality	close	Distribution
Katz centrality	Katz	Distribution
PageRank centrality	PR	Distribution

3.2   Graph Models

As mentioned above, we are mostly interested in scale-free networks, i.e., highly heterogeneous networks with many low-degree and few high-degree nodes, whose degree distribution roughly follows a power law. We thus chose Barabási-Albert, Chung-Lu, and hyperbolic random graphs as models for our experiments. Moreover, we also consider the Erdős-Rényi Graphs model, as it is arguably the most basic random graph model possible. In the following we briefly describe how graphs are generated by the different models, discuss their basic properties, and report how we did the model fitting mentioned in Sect. 2.1.

Erdős-Rényi Graphs [12]. The Erdős-Rényi random graph model is the earliest and most studied one. A graph is generated by connecting each pair of n vertices with probability p. Thus, one can control the number of vertices and expected number of edges, which is $p \cdot n(n-1)/2$. To fit the model to a given real-world network, we set n to the number of vertices and the edge probability parameter to $p = 2m/(n(n-1))$ where m denotes the number of edges in the network.

We do not expect the Erdős-Rényi Graphs model to generate very realistic graphs, i.e., we expect them to be easily distinguishable from the real-world networks.

Barabási-Albert Graphs [3]. This model (which is also called *preferential attachment*) generates a random graph by starting with a small graph of size n_0 (e.g. a cycle). Then, nodes are added one by one, each connected to k already existing nodes with probability proportional to their degree, until there are n nodes in the graph. The size of the initial graph is typically chosen as $n_0 = k$, which is the smallest value ensuring that the first node that is added in the generation process has enough neighbors to connect to. Note that $2k$ is the expected average degree of the resulting graph. Thus, to fit the model, we set n to the number of vertices and derive k from the average degree of the real-world network.

As this model generates scale-free graphs, we expect it to produce more realistic results than the Erdős-Rényi Graphs model. The main point commonly made against the Barabási-Albert model is its vanishing clustering coefficient [11], which indicates a lack of locality typically present in real-world networks.

Chung-Lu Graphs [8,9]. In the Chung-Lu model each node is assigned a weight and each pair of nodes is connected with a probability proportional to the product of their weights. In the resulting graph, each node has an expected degree equal to its weight. In our experiments, we fit the model by using the observed degree distribution in a real-world network as weights.

By construction, the Chung-Lu model mimics the degree distribution of a real-world network very well. As for the Barabási-Albert model, the most common point of criticism is its low clustering coefficient. Moreover, the Chung-Lu model seems more artificial than the Barabási-Albert model, as the latter mimics the evolution of a real-world network (in a simplified manner). The Chung-Lu model on the other hand matches the desired degree distribution much more accurately. It is thus interesting to know which of the two models leads to more realistic results with respect to features other than the degree distribution.

Hyperbolic Random Graphs [17]. In this model n nodes are placed randomly in a disk within the hyperbolic plane. Then, each pair of vertices is connected if their hyperbolic distance is below a threshold, whose size depends on n and the desired average degree. The resulting networks have a power-law degree distribution with the power-law exponent β being an input parameter. Additionally, a parameter T can be used to soften the threshold behaviour, allowing long-range edges with a certain probability. The geometry implies locality, which leads to a non-vanishing clustering coefficient [15]. Roughly speaking, the parameter T

controls how important this locality is (the more probable long-range edges are, the less important is the locality) and thus impacts the clustering coefficient.

When fitting the model parameters to a given real-world network, the largest connected component of the generated graph is typically smaller than the number of initially generated nodes n. To estimate n we use a technique based on estimating the missing nodes of degree 0 [5]. As desired average degree, we simply use the average degree of the real-world network, and the power-law exponent β is estimated based on the cumulative degree distribution. To fit the final parameter $T \in [0, 1)$, we perform a binary search on T, in each step comparing the clustering coefficients of the resulting graph and the real-world network.

Table 2. Failure rates on Facebook graphs. The table includes the same feature sets as Table 3, not showing all-0% rows. ($\varnothing$: only average values for distributions)

Feature Sets	ER	BA	CL	HRG
n, m	49%	50%	47%	44%
n, m, d	0%	0%	1%	0%
n, m, d'	1%	1%	14%	16%
$n, m, c \varnothing$	0%	0%	0%	40%
$n, m, \text{betw } \varnothing$	4%	5%	12%	41%
n, m, betw	1%	0%	7%	2%
$n, m, \text{close } \varnothing$	13%	11%	12%	30%
n, m, close	1%	3%	7%	14%
n, m, PR	0%	1%	21%	1%
$n, m, \text{Katz } \varnothing$	0%	1%	20%	2%
n, m, Katz	0%	0%	13%	0%
n, m, deg	0%	0%	42%	0%
$n, m, \text{core } \varnothing$	0%	8%	11%	2%
n, m, core	0%	0%	16%	0%
$n, m, \text{deg, betw}$	0%	0%	7%	0%
$n, m, \text{close, deg}$	0%	0%	20%	0%
$n, m, \text{PR, deg}$	0%	0%	4%	0%
$n, m, \text{Katz, deg}$	0%	0%	2%	0%
$n, m, \text{core, deg}$	0%	0%	2%	0%
$c, \beta, d \varnothing$	0%	0%	0%	1%
$c, \beta, d' \varnothing$	0%	0%	1%	19%
c, β, d'	0%	0%	1%	0%
$\text{betw, close, } d$	0%	0%	2%	1%
$\text{betw, close, } d'$	0%	1%	4%	1%

We expect hyperbolic random graphs to be more realistic than the other models, due to their non-vanishing clustering coefficient. It is, however, unclear whether hyperbolic random graphs are also more realistic with respect to other features that are not explicitly enforced by the model.

3.3   Results

The results of our experiments can be summarized as follows. We note that the insights obtained by our method are meant to guide the direction of future research rather than being reliable scientific facts by themselves.

– None of the tested models is a good representation for Facebook's social networks. Further analysis has to show if this is due to Facebook's special structure or whether it is a general issue with graphs of high average degree.
– For other real-world networks, Chung-Lu and hyperbolic random graphs outperform Erdős-Rényi and Barabási-Albert graphs, which are easy to distinguish from real-world networks even for small parameter sets. Non-surprisingly, Chung-Lu graphs perform well with respect to the degree distribution but typically have a too low clustering. Hyperbolic random graphs not only improve with respect to clustering but also outperform Chung-Lu graphs for features related to graph distances.
– Though hyperbolic random graphs have a realistic average clustering coefficient (even for the Facebook graphs), the distribution of clustering coefficients is surprisingly unrealistic.
– In the Barabási-Albert model, the choice of the initial graph is only irrelevant when the average degree is small. For networks with average degree above 30, it is easy to distinguish between graphs initialized with cliques or cycles.

Our findings are mainly based on Tables 2 and 3. They show the failure rates of the classifier trying to separate different real-world networks from generated graphs, given a subset of features. Note that a failure rate of 0% means the model can be easily distinguished while 50% means that the classifier cannot do better than guessing. To explain the resulting data, we selectively show scatter plots to visualize the relation between two parameters. The findings in this section nicely illustrate the strength of our approach: though the absolute numbers in Tables 2 and 3 have no significant meaning, their comparison can lead to interesting insights, which can be used as starting points for further investigations.

Facebook Graphs. Table 2 shows the failure rates when considering only Facebook graphs (label "socfb" in Fig. 1). Table 3 shows the same data, when excluding them. One can see that Facebook graphs are much easier to distinguish from the different models than other real-world networks: ignoring the high values for the parameters directly fitted (n and m for all models, the degree distribution for Chung-Lu and the average clustering coefficient for hyperbolic random graphs), only few parameters are well represented. Though some parameters lead to non-zero values (PageRank, average Katz centrality, and core centrality for Chung-Lu and average betweenness and closeness for hyperbolic random graphs), the failure rates are much lower than for their non-Facebook counterparts. The only exception is the average betweenness for hyperbolic random graphs.

It is interesting to note that, e.g., for the Katz centrality in Chung-Lu graphs, the failure rate drops from 20% to 2%, when additionally taking the degree distribution into account. Non-Facebook graphs behave different in this regard.

Table 3. Failure rates when excluding the Facebook graphs. (∅: only average values for distributions)

Feature Sets	ER	BA	CL	HRG
all (uncorrelated)	3%	4%	15%	14%
n, m	50%	50%	45%	46%
n, m, d	28%	22%	28%	35%
n, m, d'	37%	38%	33%	43%
n, m, c ∅	9%	9%	23%	39%
n, m, c	3%	6%	17%	29%
n, m, betw ∅	47%	49%	38%	42%
n, m, betw	2%	11%	37%	41%
n, m, close ∅	44%	47%	44%	46%
n, m, close	25%	26%	43%	45%
n, m, PR	11%	16%	43%	35%
n, m, Katz ∅	27%	27%	46%	39%
n, m, Katz	9%	15%	42%	31%
n, m, deg	5%	6%	36%	20%
n, m, core ∅	30%	46%	43%	39%
n, m, core	3%	6%	35%	19%
n, m, c, d	4%	6%	17%	28%
n, m, c, d'	4%	5%	14%	28%
n, m, c, betw	2%	2%	17%	28%
n, m, close, c	4%	5%	17%	28%
n, m, PR, c	2%	5%	18%	27%
n, m, Katz, c	1%	4%	16%	22%
n, m, core, c	1%	2%	17%	18%
n, m, deg, c	0%	2%	17%	19%
n, m, deg, betw	4%	6%	32%	21%
n, m, deg, close	6%	6%	37%	22%
n, m, deg, PR	5%	6%	32%	22%
n, m, deg, Katz	5%	7%	37%	21%
n, m, deg, core	5%	6%	30%	17%
c, β, d ∅	3%	5%	22%	43%
c, β, d	4%	5%	20%	33%
c, β, d' ∅	4%	6%	22%	36%
c, β, d'	4%	6%	20%	30%
betw, close, d	5%	6%	31%	38%
betw, close, d'	4%	6%	28%	33%

We note that most Facebook graphs have an average degree above 40, while it is below 30 for most other networks in our data set (see Fig. 3 (left)). Thus, the observed discrepancy can have its origin in the special structure of Facebook networks or in the high average degree.

Non-Facebook Graphs. The first row of Table 3 shows failure rates when using all features, partitioned into correlated groups (see Sect. 2.2). The smallest

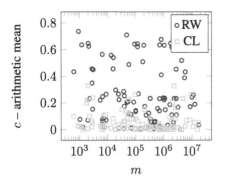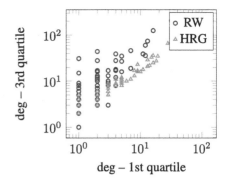

Fig. 2. Left: Average local clustering coefficients of real-world networks (not Facebook) and their Chung-Lu counterparts. Right: Degree distribution (1st/3rd quartile) of real-world networks (not Facebook) and their hyperbolic counterparts.

correlation within a group is 0.95. Erdős-Rényi and Barabási-Albert graphs are easy to distinguish from real-world networks. For Chung-Lu and hyperbolic random graphs, the classifier is wrong in about 15% of the cases.

Though the Erdős-Rényi and Barabási-Albert models perform reasonable with respect to the diameter, average betweenness, closeness, and average core centrality, they appear to be rather unrealistic in general. It is also interesting to note that the failure rate for Barabási-Albert graphs drops from 46% to 6% when considering the distribution of the core centrality instead of the average.

The main issue for Chung-Lu graphs is the clustering coefficient. For hyperbolic random graphs failure rates get worse when considering the degree distribution. Figure 2 (left) shows that for most considered networks the clustering coefficient of Chung-Lu graphs is too small. Figure 2 (right) shows that hyperbolic random graphs are not sufficiently heterogeneous: the degree of low-degree vertices is too high, while the degree of high-degree vertices is too low.

Concerning the other parameters, Chung-Lu graphs perform better than hyperbolic random graphs with respect to centrality measures that are closely related to the degree distribution (PageRank, Katz centrality, and core centrality). On the other hand, hyperbolic random graphs perform better with respect to measures related to distances (diameter, betweenness centrality, and closeness centrality). This is interesting as it supports the common claim that the metric of real-world networks is similar to the hyperbolic metric.

Distribution of the Local Clustering Coefficient. In this section, we focus on the local clustering coefficient of hyperbolic random graphs. Table 2 shows that the failure rate drops from 40% to 0% when considering the distribution instead of only the average. In Fig. 3 (left) it is easy to see that the standard deviation of the clustering coefficient is too small in hyperbolic random graphs, compared to Facebook networks. One can, however, also see, that hyperbolic random graphs can, in principal, achieve high variance. A possible explanation for the too small

standard deviation compared to Facebook graphs is that a high average degree
decreases the variance in clustering for hyperbolic random graphs.

Even though the difference becomes more apparent for graphs with high-
average degree, Fig. 3 (right) shows that hyperbolic random graphs tend to have
too little variance in the clustering coefficient. We believe that this finding provides an interesting starting point for improving the model.

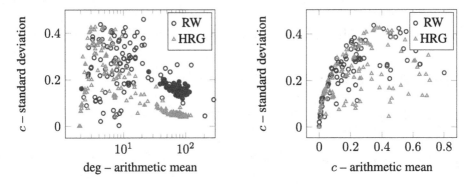

Fig. 3. Left: Standard deviation of clustering coefficients depending on the average
degree. Facebook graphs are marked with filled shapes. Right: Standard deviation of
clustering coefficients, depending on the average clustering coefficient. Facebook graphs
are excluded.

The Initial Graph in the Barabási-Albert Model. Recall that the Barabási-Albert
model generates graphs by starting with an initial graph and then successively
adding vertices, each connected to the same number of already existing vertices.
The size of the initial graph is typically chosen as small as possible, such that
the first added vertex has enough neighbors to connect to. The original paper by
Barabási and Albert [3] introducing the model does not specify how the initial
graph has to be chosen and it is generally assumed to be a negligible choice.

We compared two variants of the Barabási-Albert model using cliques and
cycles as initial graphs. For graphs with average degree at most 30, our classifier
was not able to distinguish the two different variants (50% failure rate when
using all features). If, however, the average degree is above 30, the two variants
indeed lead to graphs with different properties. Using (in addition to n and m)
the degree distribution led to an 18% failure rate, which dropped to only 5%
when using the distribution of clustering coefficients instead. Thus, for graphs
with average degree above 30, the initial graph makes a difference. This general
behaviour is not surprising as the initial graph becomes larger when increasing
the average degree. It is, however, surprising that this happens for such compar-
atively small average degrees.

4   Conclusions

We have seen that the question whether or not a machine learning technique can
successfully learn to distinguish between real-world and generated networks can

lead to interesting insights. We believe that this is particularly useful for guiding network science towards more realistic generative graph models by evaluating how good a model mimics the real world and by revealing its strengths and weaknesses.

Acknowledgements. This research has received funding from the German Research Foundation (DFG) under grant agreement no. FR 2988 (ADLON, HYP).

References

1. Attar, N., Aliakbary, S.: Classification of complex networks based on similarity of topological network features. Chaos **27**(9), 1–7 (2017)
2. Baldi, P., Brunak, S., Chauvin, Y., Andersen, C.A.F., Nielsen, H.: Assessing the accuracy of prediction algorithms for classification: an overview. Bioinformatics **16**(5), 412–424 (2000)
3. Barabási, A.L., Albert, R.: Emergence of scaling in random networks. Science **286**(5439), 509–512 (1999)
4. Bennett, K.P., Campbell, C.: Support vector machines: hype or hallelujah? SIGKDD Explor. **2**(2), 1–13 (2000)
5. Bläsius, T., Friedrich, T., Krohmer, A., Laue, S.: Efficient embedding of scale-free graphs in the hyperbolic plane. In: 24th ESA, pp. 16:1–16:18 (2016)
6. Bollobás, B., Riordan, O.: The diameter of a scale-free random graph. Combinatorica **24**(1), 5–34 (2004)
7. Bollobás, B., Riordan, O., Spencer, J., Tusnády, G.: The degree sequence of a scale-free random graph process. Random Struct. Algor. **18**(3), 279–290 (2001)
8. Chung, F., Lu, L.: The average distances in random graphs with given expected degrees. Proc. Natl. Acad. Sci. **99**(25), 15879–15882 (2002)
9. Chung, F., Lu, L.: Connected components in random graphs with given expected degree sequences. Ann. Comb. **6**(2), 125–145 (2002)
10. Easley, D., Kleinberg, J.: The small-world phenomenon. In: Networks, Crowds, and Markets: Reasoning About a Highly Connected World, Chap. 20, pp. 611–644. Cambridge University Press (2010)
11. Eggemann, N., Noble, S.D.: The clustering coefficient of a scale-free random graph. Discrete Appl. Math. **159**(10), 953–965 (2011)
12. Erdős, P., Rényi, A.: On random graphs I. Publ. Math. **6**, 290–297 (1959)
13. Fernández-Delgado, M., Cernadas, E., Barro, S., Amorim, D., Amorim Fernández-Delgado, D.: Do we need hundreds of classifiers to solve real world classification problems? J. Mach. Learn. Res. **15**, 3133–3181 (2014)
14. Friedrich, T., Krohmer, A.: On the diameter of hyperbolic random graphs. In: Halldórsson, M.M., Iwama, K., Kobayashi, N., Speckmann, B. (eds.) ICALP 2015. LNCS, vol. 9135, pp. 614–625. Springer, Heidelberg (2015). https://doi.org/10.1007/978-3-662-47666-6_49
15. Gugelmann, L., Panagiotou, K., Peter, U.: Random hyperbolic graphs: degree sequence and clustering. In: Czumaj, A., Mehlhorn, K., Pitts, A., Wattenhofer, R. (eds.) ICALP 2012. LNCS, vol. 7392, pp. 573–585. Springer, Heidelberg (2012). https://doi.org/10.1007/978-3-642-31585-5_51
16. Karp, R.M.: The probabilistic analysis of combinatorial optimization algorithms. In: Proceedings of the International Congress of Mathematicians, pp. 1601–1609 (1983)

17. Krioukov, D., Papadopoulos, F., Kitsak, M., Vahdat, A., Boguñá, M.: Hyperbolic geometry of complex networks. Phys. Rev. E **82**(3), 036106 (2010)
18. Rossi, R.A., Ahmed, N.K.: The network data repository with interactive graph analytics and visualization. In: Proceedings of the Twenty-Ninth AAAI Conference on Artificial Intelligence (2015). http://networkrepository.com
19. Schölkopf, B., Burges, C.J.C., Smola, A.J. (eds.): Advances in Kernel Methods: Support Vector Learning. MIT Press, Cambridge (1999)
20. Soundarajan, S., Eliassi-Rad, T., Gallagher, B.: A guide to selecting a network similarity method. In: SDM, pp. 1037–1045 (2014)
21. Staudt, C.L., Sazonovs, A., Meyerhenke, H.: NetworKit: a tool suite for large-scale complex network analysis. Netw. Sci. **4**(4), 508–530 (2016)
22. Watts, D.J.: A simple model of global cascades on random networks. Proc. Natl. Acad. Sci. **99**(9), 5766–5771 (2002)
23. Watts, D.J., Strogatz, S.H.: Collective dynamics of "small-world" networks. Nature **393**, 440–442 (1998)

Dynamic Competition Networks: Detecting Alliances and Leaders

Anthony Bonato[1(✉)], Nicole Eikmeier[2], David F. Gleich[2], and Rehan Malik[1]

[1] Ryerson University, Toronto, Canada
abonato@ryerson.ca
[2] Purdue University, Lafayette, USA

Abstract. We consider social networks of competing agents that evolve dynamically over time. Such dynamic competition networks are directed, where a directed edge from nodes u to v corresponds a negative social interaction. We present a novel hypothesis that serves as a predictive tool to uncover alliances and leaders within dynamic competition networks. Our focus is in the present study is to validate it on competitive networks arising from social game shows such as Survivor and Big Brother.

1 Introduction

Complex social networks are heterogeneous, evolving, and pervasive in the natural world and in technological settings. Social networks present rich sources of complex networks, where nodes represent agents and edges correspond to some form of social interaction. For example, in Facebook edges represent friendship, while on Twitter they denote following. Complex, social networks commonly display power law degree distributions, the small world property (short distances between nodes and high local clustering) and other phenomena such as densification and strong community structure; see [4,8,10]. Another key principle underlying social networks is that links exhibit homophily, that is, nodes with similar social attributes are linked, which is related to an embedding of the nodes in a so-called *Blau space*, where nodes are assigned to points in a suitable metric space and the relative distance between pairs of nodes is a function of similar social attributes. See [5,17].

While social interaction is usually studied from the premise of friendship, cooperation, or other positive social interactions, there is a growing literature on the study of *negative* social interaction as a generative mechanism underlying social networks. For example, while transitivity is a folkloric notion in social networks, summarized in the adage that "friends of friends are more likely friends," structural balance theory (see [10,14] for a modern treatment) points also to the inverse adage "enemies of enemies are more likely friends." A common problem in this direction is the prediction of the type of edges in a social

Research supported by grants from NSERC, Ryerson University, NSF CCF-1149756, IIS-1546488, NSF Center for Science of Information, CCF-0939370, DARPA SIMPLEX.

network [16, 19, 21]. Hence, competitive and negative relationships are critically important to the study of social networks, and are often hidden drivers of link formation.

Competitive relationships were studied recently via the Iterated Local Anti-Transitivity (or ILAT) model; see [6, 7]. In the ILAT model, each node u duplicates every time-step by forming its *anti-clone* u', so that u' joins to the nodes in the non-neighbor set of u. We may also consider real-world networks of opposing nation states, rival gangs or other organizations, and consider alliances formed by mutually shared adversaries. The ILAT model provably generates highly dense graphs with low diameter and high local clustering. See [13] for a recent study using the spatial location of cities to form an interaction network, where links enable the flow of cultural influence, and may be used to predict the rise of conflicts and violence. Another example comes from market graphs, where the nodes are stocks, and stocks are adjacent as a function of their correlation measured by a threshold value $\theta \in (0, 1)$. Market graphs were considered in the case of negatively correlated (or competitive) stocks, where stocks are adjacent if $\theta < \alpha$, for some positive α; see [3].

In the present paper, we focus on the underlying structure of social networks of competitors that evolve dynamically over time. We view such networks as directed, where a directed edge from nodes u to v corresponds to some kind of negative social interaction. For example, a directed edge may represent a vote by one player for another in a social game such as the television program Survivor. Directed edges are added over discrete time-steps in what we call dynamic competitive networks. Our main contribution in this empirical work is a hypothesis that serves as a predictive tool to uncover alliances and leaders within dynamic competition networks. While the hypothesis may hold more broadly, our focus here is on competitive networks arising from social game shows. We validate the hypothesis using voting record data of the social game shows Survivor and Big Brother.

We organize the discussion in this paper as follows. In Sect. 2, we formally introduce dynamic competition networks, and using graph theoretic tools, give a precise formulation of the Dynamic Competition Hypothesis. In Sect. 3 and the Appendix, we present voting data from all the seasons of U.S. Survivor and Big Brother, focusing on three seasons of Survivor in detail and one season of Big Brother. We analyze this data using tools from network science in an effort to validate the Dynamic Competition Hypothesis. We find that the hypothesis accurately predicts the emergence of alliances and predicts finalists with a high degree of precision. The final section interprets our results within the context of real-world complex networks, and presents open problems derived from our analysis.

We consider directed graphs with multiple directed edges throughout the paper. For background on graph theory, the reader is directed to [20]. Additional background on complex networks may be found in the book [4].

2   Dynamic Competition Hypothesis

A *competition network* G is one where nodes represent agents, and there is directed edge between nodes u and v in G if agent u is in competition with agent v. The directed edge (u, v) may also represent a vote against v (depending on the nature of G). A *dynamic competition network* is a competition network where directed edges are added over discrete time-steps. For example, on the game show Survivor (as we discuss in detail in the next section), players cast votes against each other, and the votes correspond to directed edges in the network. As another example, nodes may consist of nation states and edges correspond to conflicts between them. Dynamic competition networks may have multiple edges. Note that dynamic competition networks are also models of (sports) tournaments. However, in dynamic competition networks, not all nodes are joined by edges as is typically the case in tournaments. Our focus in this work will be on dynamic competition networks arising in social networks, and we focus specifically on networks arising from Survivor and Big Brother.

Before we describe our hypothesis about the structure of competition networks, we present some graph-theoretic terminology. We consider standard metrics in network science, such as in- and out-degree, closeness and betweenness. Given the nature of the voting network in Survivor, we also consider the number of common out-neighbors as a key metric.

For nodes u, v, and w, we say that w is a *common out-neighbor* of u and v if (u, w) and (v, w) are directed edges. For a pair of distinct nodes u, v, we define $\mathrm{CON}(u, v)$ to be the number of common out-neighbors of u and v. For a fixed node u, define

$$\mathrm{CON}(u) = \sum_{v \in V(G)} \mathrm{CON}(u, v).$$

We call $\mathrm{CON}(u)$ the *CON score* of u. For a set of vertices S with at least two nodes, we define

$$\mathrm{CON}(S) = \sum_{u,v \in S} \mathrm{CON}(u, v).$$

Note that $\mathrm{CON}(S)$ is a non-negative integer.

A set of nodes S with no directed edges in its induced subgraph is called *independent*; we also need a notion of being "close" to independent. For a set S of nodes, define its *edge density* to be the ratio $ED(S) = |E(S)|/\binom{|S|}{2}$. Observe that $ED(S)$ may be greater than 1 as there may be multiple edges in the digraphs we consider. For a non-negative real number ϵ say that a set S is *ϵ-near independent* if $ED(S) \le \epsilon$. The parameter ϵ measures the relative density of sets of vertices. We say that a set is near independent if it is *ϵ-near independent* for some positive value of ϵ; typically, in applications, we take ϵ to be small. The value of ϵ will often be heuristically determined in a real-world networks by considering a ranking of subsets by their edge density. Note that independent sets are trivially near independent.

For a strongly connected digraph G and a node v, define the *closeness* of u by

$$C(u) = \left(\sum_{v \in V(G) \backslash \{u\}} d(u, v) \right)^{-1}$$

where $d(u, v)$ corresponds to the distance measured by one-way, directed paths from u to v. The *betweenness* of v is defined by

$$B(v) = \sum_{x, y \in V(G) \backslash \{v\}} \sigma_{xy}(v)/\sigma_{xy},$$

where $\sigma_{xy}(v)$ is the number of shortest one-way, directed paths between x and y that go through v, and σ_{xy} is the number of shortest one-way, oriented paths between x and y. Both closeness and betweenness are well-studied centrality measures for complex networks [9]. For example, centrality of sports networks is often used to rank teams [15].

2.1   The Hypothesis

Alliances are defined as groups of agents who pool capital towards mutual goals. In the context of social game shows such as Survivor, alliances are groups of players who work together to vote off players outside the alliance. Members of an alliance are typically less likely to vote for each other, and this is the case in strong alliances. *Leaders* are defined as members with high standing in the network, and edges emanating from leaders may influence edge creation in other agents. In Survivor, leaders may be the winner of a given season, but may also be non-winning players with a strong influence on the outcomes of the game. One of our main goals is to apply network science to help determine alliances and leaders in dynamic competitive networks arising in social networks.

The *Dynamic Competition Hypothesis* (or *DCH*) asserts that dynamic competition networks arising from a social networks satisfy the following four properties.

1. Alliances are near independent sets.
2. Strong alliances have low edge density.
3. Members of an alliance with high CON scores are more likely leaders.
4. Leaders exhibit high closeness, high CON scores, low in-degree, and high out-degree.

The DCH provides a quantitative framework for the structure of dynamic competition networks arising from social networks; no other data is required other than the presence of competitive relationships. See Fig. 1 for a visualization of the DCH.

Note how items (1), (2), and (3) mutually reinforce each other. Once we have discovered an alliance as per (1), we can measure its strength relative to other alliances via (2), and use (3) as tool to isolate leaders within alliances.

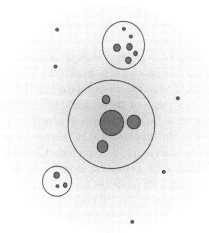

Fig. 1. A heat map representation of dynamic competition networks according to the DCH, where nodes closer to the center have higher closeness and CON scores. Larger nodes have higher CON scores, lower in-degree, and higher out-degree. The subsets correspond to alliances.

Item (4) is independent of alliances; in particular, while we expect leaders to be in alliances (that is, have prominent local influence), leaders are determined via global metrics of the network.

Interestingly, closeness rather than betweenness appears be a good centrality measure in the dynamic competition networks studied in the next section. This may be explained by the low in-degree of nodes corresponding to leaders.

3   Data and Methods

We extracted data from the American television series Survivor over all of its seasons, and for further validation, from all seasons of Big Brother. Before we present the data in detail for a subset of seasons, we give some background on both series. Survivor and Big Brother are examples of social games, where social interactions help determine the gameplay and winner. We focus on the US version of both shows, but they play in several countries, accounting for over one hundred seasons in total.

In Survivor, strangers called *survivors* are placed in a location and forced to provide shelter and food for themselves, with limited support from the outside world. Survivors are split into two or more *tribes* which cohabitate and work together. Tribes compete for immunity and the losing tribe goes to tribal council where one of their members is voted off. At some point during the season, tribes merge and the remaining survivors compete for individual immunity. Survivors voted off may be part of the *jury*. When there are a small number of remaining survivors who are *finalists* (typically two or three), the jury votes in favor of one of them to become the *Sole Survivor* who receives a cash prize of one million dollars.

In Big Brother, a group of strangers called *HouseGuests* cohabitate in a custom set under video surveillance. Each week, the HouseGuests compete for the title of *Head of Household*, who must nominate two HouseGuests for eviction. The Houseguests vote to evict one of them, and the one with the most votes is evicted. The winner received a cash prize of half a million dollars.

In both Survivor and Big Brother, several twists have been introduced during the seasons. For example, in Survivor, these include the introduction of a hidden immunity idol which would protect a survivor from being voted out if used during tribal council. As a disclaimer, our analysis is insensitive to these twists.

Data was taken from Survivor Wiki [18] and Big Brother Wiki [2], which contains information on contestants, their voting records and tribes, and catalogues of alliances. For computing centrality metrics and for the dynamic competition graph visualization, we used the open source Gephi software [1].

We present below visualizations of the dynamic competition networks for Survivor: Borneo, China, Game Changes, and HHH; we also include data from Season 12 of Big Brother. Note that the data is taken after all votes had been cast against other players, and tables are provided with a summary of relevant network statistics. The order of the tables is given by their elimination order from the game, so the first entry is the winner and the others are ordered by when they were eliminated. In all of the five seasons described below, the data conforms to the predictions of the DCH with regards to leaders (that is, winners in this context). It also clearly delineates alliances, as we discuss below.

3.1 Borneo

We consider the first season of Survivor set in Borneo. The abbreviations ID, OD, C, CON, and B stand for in-degree, out-degree, closeness, CON-score, and betweenness, respectively.

Note that Richard, the Sole Survivor of the season, has one of the highest closeness and CON scores. Rudy and Susan have higher scores, however. We note that Kelly won individual immunity several times near the end of the game, and her voting out Rudy and Susan was a deciding factor in Richard's win. We also note that comparing betweenness of players is inconclusive as a predictor of leaders. For example, we computed Richard's betweenness as 28.7, Kelly's as 0, and Rudy's as 36.5. One explanation of this is that leaders tend to have lower in-degree, which may reduce the number of paths traversing through them. As such, we do not include betweenness scores for other seasons.

Name	Tribe	ID	OD	C	CON	B
Richard	Tagi	6	10	0.737	42	28.7
Kelly	Tagi	0	12	0.682	34	0
Rudy	Tagi	8	11	0.778	45	36.483
Susan	Tagi	7	10	0.778	44	16.467
Sean	Tagi	9	9	0.7	38	17.917
Colleen	Pagong	7	8	0.636	29	33.067
Gervaise	Pagong	6	7	0.636	31	8.583
Jenna	Pagong	11	6	0.583	27	27.85
Greg	Pagong	6	5	0.412	15	4.833
Gretchen	Pagong	4	4	0.56	17	7.233
Joel	Pagong	4	3	0.412	17	1
Dirk	Tagi	4	3	0.5	12	1.317
Ramona	Pagong	6	2	0.412	10	17.733
Stacey	Tagi	6	2	0.452	4	1.733
B.B	Pagong	6	1	0.298	5	0.333
Sonja	Tagi	4	1	0.452	4	0.75

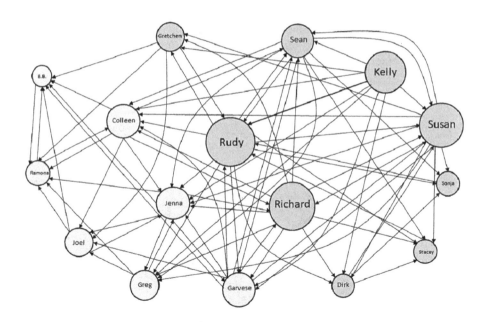

3.2 China

We next turn to Survivor: China, which was chosen because it represents a sample after the game was better known, and contestants better understood which strategies to employ in the game.

Name	Tribe	ID	OD	C	CON
Todd	Fei Long	5	9	0.765	49
Courtney	Fei Long	0	9	0.667	39
Amanda	Fei Long	0	9	0.737	49
Denise	Fei Long	3	9	0.722	40
Peih-Gee	Zhan Hu	8	10	0.722	41
Erik	Zhan Hu	5	9	0.722	41
James	Fei Long	9	6	0.591	31
Frosti	Zhan Hu	7	7	0.65	39
Jean-Robert	Fei Long	12	4	0.5	23
Jaime	Zhan Hu	7	5	0.481	26
Sherea	Zhan Hu	6	4	0.448	24
Aaron	Fei Long	3	2	0.406	12
Dave	Zhan Hu	6	3	0.382	11
Leslie	Fei Long	6	1	0.342	9
Ashley	Zhan Hu	8	2	0.464	10
Chicken	Zhan Hu	5	1	0.333	6

In this season, it is evident that Todd, the Sole Survivor, is the clear front-runner for Sole Survivor based on his high closeness and CON scores. Courtney and Amanda emerge also as leaders based on their scores.

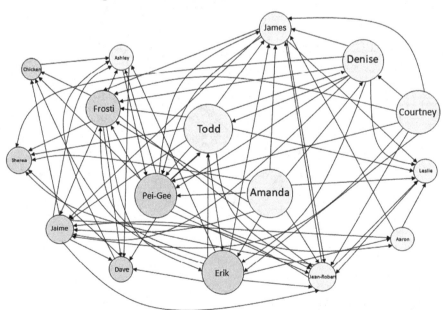

3.3   Game Changers

We next analyzed Survivor: Game Changers, as the second-to-last season of the show.

Name	Tribe	ID	OD	C	CON
Sarah	Nuku	3	13	0.692	64
Brad	Nuku	2	12	0.643	49
Troyzan	Mana	2	12	0.643	55
Tai	Nuku	12	13	0.72	56
Aubry	Mana	9	13	0.72	61
Cirie	Nuku	0	8	0.613	45
Michaela	Mana	11	11	0.643	51
Andrea	Nuku	14	8	0.581	39
Sierra	Nuku	15	7	0.581	34
Zeke	Nuku	11	6	0.6	39
Debbie	Nuku	6	7	0.545	32
Ozzy	Nuku	7	4	0.5	22
Hali	Mana	8	5	0.474	28
Jeff	Mana	6	5	0.529	33
Sandra	Mana	5	5	0.581	34
JT	Nuku	3	2	0.45	18
Malcom	Mana	5	3	0.439	24
Caleb	Mana	5	3	0.4	21
Tony	Mana	7	2	0.439	15
Ciera	Mana	9	1	0.4	8

In this season, the Sole Survivor Sarah has high closeness and CON scores, but Tai and Aubry have higher closeness scores. Note, however, both players have high in-degrees which likely disadvantaged them.

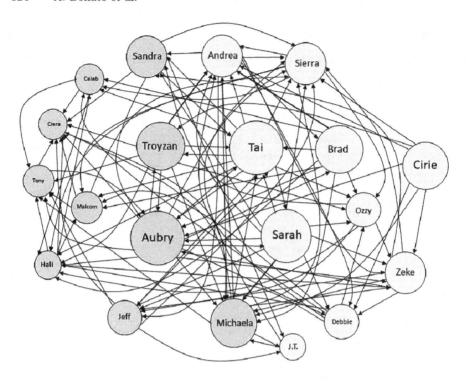

3.4   HHH

We now turn to the most recent season of Survivor, Survivor: Heroes vs Healers vs Hustlers (or HHH, for short). The following table contains network data for Survivor: HHH.

Name	Tribe	ID	OD	C	CON
Ben	Levu	11	11	0.63	41
Chrissy	Levu	7	13	0.68	44
Ryan	Yawa	2	14	0.708	47
Devon	Yawa	2	11	0.708	55
Mike	Soko	9	9	0.63	37
Ashley	Levu	8	10	0.607	46
Lauren	Yawa	3	7	0.63	39
Joe	Soko	12	6	0.607	26
JP	Levu	6	8	0.586	25
Cole	Soko	7	4	0.531	26
Desi	Soko	11	3	0.515	9
Jessica	Soko	7	1	0.415	6
Ali	Yawa	3	4	0.5	19
Roark	Soko	3	1	0.415	6
Alan	Levu	2	2	0.415	11
Patrick	Yawa	5	2	0.405	6
Simone	Yawa	5	1	0.293	4
Katrina	Levu	5	1	0.386	5

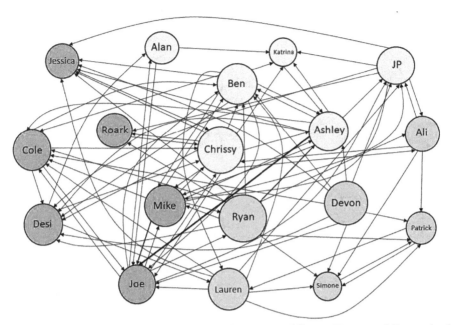

The finalists of this season were Ben, Chrissy and Ryan. Ryan and Devon had the highest overall closeness and highest overall CON scores, followed by Chrissy. However, Ben, the Sole Survivor, had lower scores than the other finalists; he secured his place in the final three by playing the hidden immunity idol three times.

3.5 Big Brother

Given the success of the DCH in Survivor, we turned to data from another social game Big Brother, focusing on Season 12.

Name	ID	OD	C	CON
Hayden	3	16	0.923	44
Lane	3	10	0.857	46
Enzo	4	9	0.8	48
Britney	4	10	0.8	43
Regan	5	8	0.706	49
Brendon	7	9	0.706	40
Matt	9	7	0.632	35
Kathy	7	4	0.6	20
Rachel	8	6	0.667	24
Kristen	7	3	1	25
Andrew	9	2	1	17
Monet	8	1	1	10
Annie	11	0	0	0

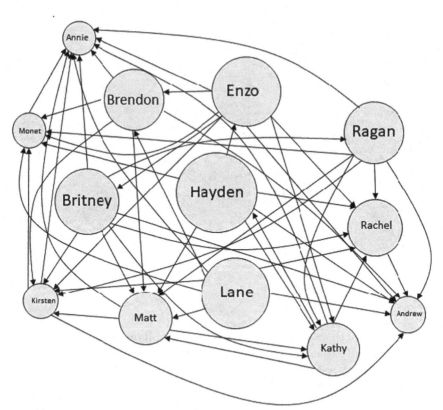

Hayden, the winner of the season, is the clear frontrunner with regards to closeness and CON scores, with HouseGuests Lane and Enzo rounding out the top three.

3.6   CON Scores

The CON score for each player in the full set of Survivor and Big Brother seasons is listed in the Appendix; there are 35 seasons in Survivor and 20 for Big Brother. In this section, we summarize that data. We are interested in knowing if a high CON score correlates with being the winner. To test this, we check whether the winner of a particular season has a CON score within the top three or five CON scores out of every player from that season. As displayed in the table below, 68.6% of winners in Survivor had a top three CON score, and 94.3% of them have a top five CON score.

We compare the CON score to two other well-known rankings: PageRank and Jaccard Similarity scores. Jaccard Similarity is a type of normalized CON score, and both of these methods are commonly used in ranking; see, for example, [11,12]. Note that we computed PageRank scores on the *reverse* of the network discussed in Sect. 2. The table shows that the CON scores are the best predictor for winners in Survivor, while PageRank is a slightly better predictor in Big Brother. Furthermore, we calculate the probability of the winner appearing in a random set of three or five, under the *random set* column. This probability varies depending on the size of the network (that is, the number of players). We see, for example, that the probability of a winner being in a random set of three in Survivor is between 15% and 18.8%. In all cases, these probabilities are lower than the CON scores, which suggests that the result of the winner having one of the largest CON scores is not due to random chance.

		CON	Page rank	Jacard similarity	Random set
Survivor	**Top 3**	68.6	54.3	54.3	15.0–18.8
	Top 5	94.3	88.6	80.0	25.0–31.3
Big Brother	**Top 3**	60.0	80.0	25.0	17.6–30.0
	Top 5	70.0	100	55.0	29.4–50.0

3.7 Alliances

In addition to predicting winners, we analyzed alliances in the various seasons and computed their edge density. All the alliances conform to the DCH as they form near independent sets. Some alliances have relatively high edge density, as we note in the Tagi alliance in Borneo (which includes the sole survivor Richard). Nevertheless, narrowing down the alliances to subsets of finalists appears to reduce the edge density. For example, in the Tagi alliance, the edge density of the subsets {Kelly,Richard} is 1/2 and {Richard,Rudy} is 0. Analogously, in the Fie-Long alliance in Survivor: China, the subset {Amanda, Courtney, Todd} has edge density 0.

Season	Winner	Finalists	Alliances	ED
Borneo	Richard	Kelly	*Barbecue*: Colleen, Jenna, Gervase	1.667
			Tagi: Richard, Rudy, Susan, Kelly	1.5
China	Todd	Courtney	*Fei Long*: Todd, Courtney, Amanda, Aaron, Denise, James, Frosti	0.667
		Amanda	*Zhan Hu*: Peih-Gee, Erik, Jaime	0.0
Game Changers	Sarah	Brad	*Power Six*: Sarah, Brad, Troyzan, Sierra, Debbie, Tai	0.933
		Troyzan	*Tavua*: Aubry, Cirie, Michaela, Ozzy, Andrea, Zeke, Sarah	1.238
HHH	Ben	Chrissy	*Healers*: Joe, Desi, Jessica, Cole, Mike	0.6
		Ryan	*The Round Table*: Chrissy, Ryan, Devon, JP, Ben, Ashley, Lauren	0.905
			Final Four: Ashley, Lauren, Ben, Devon	1.333
Big Brother 12	Hayden	Lane	*The Brigade*: Enzo, Hayden, Lane, Matt	0.5

We list the edge densities for each alliance in the Appendix, along with the edge density for the entire graph. There may be some use in exploring to what extent alliances have smaller edge density than that of the entire graph. As already discussed, the edge density of an alliance can become much lower when removing players who play against their alliance. That being said, 60% of the Survivor seasons have an alliance with a lower edge density than the edge density for the total graph, and 95% of Big Brother seasons have an alliance with a lower edge density than the edge density for the total graph. More exploration is needed to understand the relationship between the edge densities of alliances and leaders.

4   Discussion and Future Work

We introduced the notion of dynamic competition networks and studied their properties. The Dynamic Competition Hypothesis (DCH) was presented, which resolves dynamic competition networks arising from social networks into alliances, detects leaders, and measures the relative strength of alliances. The DCH was tested with voting data from all seasons of the U.S. television social game shows Survivor and Big Brother. In all seasons and as predicted by the DCH, alliances correspond to near independent sets, CON scores accurately determine leaders of alliances, and leaders are detected via their CON scores and closeness.

In future work, we will mine data from all international seasons of Survivor and Big Brother (our current analysis uses only seasons from a single country). We will also look for other data sets to further validate the DCH more broadly, within the lens of structural balance theory and social network analysis. A weakness of our current theory is that longer lasting members of a season accumulate more influence simply due to their survival. In particular, players in Survivor and Big Brother that survive longer in the game have a greater opportunity to improve their CON-scores and other metrics. In future work, we will therefore, evaluate data at earlier stages of the formation of the network. Other areas where we can explore the DCH are food webs, signed networks (by extracting the subgraph with negative signs), and geo-political networks. It would be interesting to invert the DCH to determine low ranked members of dynamic competition networks. Further, it would be useful to develop a mathematical model predicting the evolution of dynamic competition networks, which provably simulates properties predicted by the DCH.

A   Appendix

Complete data from all U.S. Seasons of Survivor and Big Brother may be found in the document:

http://www.math.ryerson.ca/~abonato/papers/SurvivorBB_Data_BEGM.

We include this data below for convenience. The data from five of these seasons (Survivor: Borneo, China, Game Changers, HHH, and Big Brother 12) is discussed in detail in the body of the paper. We provide the remaining data to further support the DCH and for transparency.

All data was gathered from the Survivor and Big Brother Wiki pages [2,18]. Within the wiki, each season has a dedicated page (for example http://survivor. wikia.com/wiki/Survivor:_Millennials_vs._Gen_X) with a table of voting history, which was used to construct the directed networks. Each player of the game is a vertex of the network, with a directed edge added from vertex A to vertex B if player A voted against player B. If player A voted against player B n times, then the edge has a weight of n. We scraped voting history tables using simple python code, and further did the analysis in python.

Appendix A.1 gives a table for each season with the following metrics, which are discussed thoroughly in Sect. 2:

1. In-degree;
2. Out-degree;
3. Closeness;
4. CON Score.

Contestants are listed in the order which they were voted out, where the player on top of the table remained in the game the longest.

Appendix A.2 gives the edge density for every Alliance in each season, as well as the edge density of the full graph for comparison. Information on edge density can be found in Sect. 2.

A.1   Complete Network Metrics

Africa			
Name	ID	OD	C	CON
Ethan	0	10	0.75	51
Kim J.	1	11	0.824	57
Lex	10	11	0.737	47
Tom	9	10	0.7	43
Teresa	4	10	0.636	35
Kim P.	4	10	0.667	39
Frank	9	8	0.636	31
Brandon	6	8	0.636	32
Kelly	5	6	0.56	30
Clarence	12	4	0.538	20
Lindsey	12	3	0.5	12
Silas	8	4	0.452	13
Linda	4	3	0.378	11
Carl	7	1	0.341	6
Jessie	5	2	0.368	14
Diane	6	1	0.359	9

All-Stars			
Name	ID	OD	C	CON
Amber	6	8	0.682	36
Rob M.	1	8	0.682	34
Jenna L.	4	8	0.682	32
Rupert	4	8	0.682	32
Tom	4	6	0.6	28
Shii Ann	5	8	0.625	33
Alicia	7	4	0.536	18
Kathy	6	5	0.556	18
Lex	7	5	0.577	21
Jerri	7	6	0.6	21
Ethan	6	5	0.536	15
Colby	4	2	0.395	10
Susan	0	1	0.276	4
Richard	6	1	0.288	3
Rob C.	5	1	0.357	6
Jenna M.	0	0	0	0
Rudy	3	2	0.375	8
Tina	4	1	0.417	3

Blood vs. Water			
Name	ID	OD	C	CON
Tyson	2	12	0.63	49
Monica	6	13	0.708	67
Gervase	6	12	0.654	54
Tina	10	9	0.654	54
Ciera	14	14	0.739	62
Hayden	8	13	0.739	75
Katie	4	15	0.68	64
Caleb	4	11	0.68	74
Laura M.	19	6	0.567	38
Vytas	10	10	0.68	58
Aras	7	3	0.436	16
Laura B.	11	4	0.5	34
Kat	5	3	0.486	27
Brad	7	5	0.531	36
John	8	4	0.515	31
Colton	0	1	0.281	5
Rachel	5	3	0.5	21
Marissa	10	2	0.436	7
Rupert	0	1	0.375	13
Candice	6	1	0.37	13

Borneo			
Name	ID	OD	C	CON
Richard	6	10	0.737	42
Kelly	0	12	0.682	34
Rudy	8	11	0.778	45
Susan	7	10	0.778	44
Sean	9	9	0.7	38
Colleen	7	8	0.636	29
Gervase	6	7	0.636	31
Jenna	11	6	0.583	27
Greg	6	5	0.412	15
Gretchen	4	4	0.56	17
Joel	4	3	0.412	17
Dirk	4	3	0.5	12
Ramona	6	2	0.412	10
Stacey	6	2	0.452	4
B.B.	6	1	0.298	5
Sonja	4	1	0.452	4

Cagayan			
Name	ID	OD	C	CON
Tony	5	9	0.696	44
Woo	4	10	0.696	42
Kass	2	12	0.696	41
Spencer	8	12	0.667	34
Trish	5	8	0.64	29
Tasha	4	10	0.696	36
Jefra	9	8	0.552	28
Jeremiah	9	7	0.571	32
LJ	8	6	0.516	23
Morgan	8	4	0.471	15
Sarah	6	2	0.381	12
Alexis	8	2	0.381	8
Lindsey	0	1	0.354	6
Cliff	4	1	0.348	6
J'Tia	7	3	0.444	11
Brice	5	1	0.281	6
Garrett	3	2	0.32	7
David	4	1	0.314	4

Cambodia			
Name	ID	OD	C	CON
Jeremy	3	12	0.643	45
Spencer	11	15	0.72	61
Tasha	8	13	0.692	53
Kelley	17	13	0.692	46
Keith	4	10	0.6	31
Kimmi	5	10	0.621	43
Abi-Maria	14	12	0.72	48
Joe	8	5	0.545	29
Stephen	9	7	0.621	43
Ciera	10	5	0.514	16
Kelly	8	7	0.563	35
Andrew	4	5	0.563	26
Kass	6	2	0.462	8
Woo	5	5	0.514	24
Terry	0	2	0.487	17
Monica	3	1	0.383	6
Jeff	4	4	0.383	16
Peih-Gee	4	3	0.462	19
Shirin	5	2	0.439	12
Vytas	6	1	0.429	10

Caramoan			
Name	ID	OD	C	CON
Cochran	0	14	0.655	52
Dawn	2	14	0.692	57
Sherri	6	13	0.692	59
Eddie	16	12	0.667	37
Erik	2	10	0.621	47
Brenda	5	10	0.621	48
Andrea	13	8	0.643	49
Reynold	9	9	0.581	25
Malcolm	11	6	0.581	33
Phillip	5	8	0.563	40
Michael	10	8	0.6	39
Corinne	7	6	0.545	24
Julia	9	6	0.545	22
Matt	4	5	0.529	27
Brandon	8	2	0.439	12
Laura	6	4	0.409	14
Shamar	8	2	0.429	13
Hope	8	2	0.316	3
Allie	6	1	0.305	3
Francesca	6	1	0.4	8

China			
Name	ID	OD	C	CON
Todd	5	9	0.765	49
Courtney	0	9	0.667	39
Amanda	0	9	0.737	49
Denise	3	9	0.722	40
Peih-Gee	8	10	0.722	41
Erik	5	9	0.722	41
James	9	6	0.591	31
Frosti	7	7	0.65	39
Jean-Robert	12	4	0.5	23
Jaime	7	5	0.481	26
Sherea	6	4	0.448	24
Aaron	3	2	0.406	12
Dave	6	3	0.382	11
Leslie	6	1	0.342	9
Ashley	8	2	0.464	10
Chicken	5	1	0.333	6

Cook Islands			
Name	ID	OD	C	CON
Yul	5	9	0.633	46
Ozzy	1	10	0.633	42
Becky	5	9	0.633	46
Sundra	6	10	0.594	41
Adam	5	11	0.633	52
Parvati	4	10	0.633	52
Jonathan	15	9	0.613	43
Candice	6	7	0.543	40
Nate	5	7	0.543	37
Jenny	6	6	0.5	39
Rebecca	6	6	0.528	36
Brad	7	4	0.475	29
Jessica	6	3	0.432	15
Cristina	5	4	0.422	21
Cao Boi	6	2	0.422	9
Stephannie	9	3	0.38	12
J.P.	7	2	0.352	10
Cecilia	5	2	0.432	5
Billy	4	1	0.396	0
Sekou	3	1	0.38	5

Fiji			
Name	ID	OD	C	CON
Earl	1	9	0.654	36
Cassandra	5	7	0.586	28
Dreamz	2	11	0.654	42
Yau-Man	9	10	0.68	39
Boo	5	6	0.567	21
Stacy	4	6	0.567	26
Alex	9	8	0.63	37
Mookie	6	10	0.654	38
Edgardo	5	5	0.531	25
Michelle	3	5	0.486	24
Lisi	7	3	0.472	10
Rocky	5	6	0.531	26
Anthony	10	5	0.425	18
Rita	6	4	0.472	17
Liliana	6	1	0.386	4
Gary	0	0	0	0
Sylvia	6	2	0.37	12
Erica	6	2	0.436	10
Jessica	6	1	0.327	5

Gabon			
Name	ID	OD	C	CON
Bob	2	9	0.615	31
Susie	7	10	0.727	44
Sugar	0	10	0.708	49
Matty	7	12	0.762	54
Ken	7	12	0.762	52
Crystal	11	11	0.762	50
Corinne	4	5	0.593	26
Randy	5	5	0.552	27
Charlie	5	3	0.516	17
Marcus	3	3	0.552	16
Dan	4	3	0.457	19
Ace	5	5	0.5	25
Kelly	8	4	0.5	21
G.C.	6	4	0.421	24
Jacquie	5	2	0.356	12
Paloma	7	1	0.34	4
Gillian	8	2	0.457	12
Michelle	8	1	0.32	7

Game Changers			
Name	ID	OD	C	CON
Sarah	3	13	0.692	64
Brad	2	12	0.643	49
Troyzan	2	12	0.643	55
Tai	12	13	0.72	56
Aubry	9	13	0.72	61
Cirie	0	8	0.613	45
Michaela	11	11	0.643	51
Andrea	14	8	0.581	39
Sierra	15	7	0.581	34
Zeke	11	6	0.6	39
Debbie	6	7	0.545	32
Ozzy	7	4	0.5	22
Hali	8	5	0.474	28
Jeff	6	5	0.529	33
Sandra	5	5	0.581	34
J.T.	3	2	0.45	18
Malcolm	5	3	0.439	24
Caleb	5	3	0.4	21
Tony	7	2	0.439	15
Ciera	9	1	0.4	8

Guatemala			
Name	ID	OD	C	CON
Danni	1	12	0.739	50
Stephenie	2	11	0.739	64
Rafe	2	11	0.739	64
Lydia	10	11	0.739	56
Cindy	6	9	0.654	47
Judd	5	8	0.607	43
Gary	8	9	0.567	36
Jamie	10	7	0.586	41
Bobby Jon	8	6	0.567	27
Brandon	6	5	0.447	22
Amy	4	5	0.447	29
Brian	7	4	0.436	24
Margaret	7	3	0.472	19
Blake	5	2	0.34	12
Brooke	5	2	0.459	15
Brianna	7	2	0.447	15
Morgan	8	1	0.436	8
Jim	8	1	0.327	6

Heroes vs. Healers vs. Hustlers			
Name	ID	OD	C	CON
Ben	11	11	0.63	41
Chrissy	7	13	0.68	44
Ryan	2	14	0.708	47
Devon	2	11	0.708	55
Mike	9	9	0.63	37
Ashley	8	10	0.607	46
Lauren	3	7	0.63	39
Joe	12	6	0.607	26
JP	6	8	0.586	25
Cole	7	4	0.531	26
Desi	11	3	0.515	9
Jessica	7	1	0.415	6
Ali	3	4	0.5	19
Roark	3	1	0.415	6
Alan	2	2	0.415	11
Patrick	5	2	0.405	6
Simone	5	1	0.293	4
Katrina	5	1	0.386	5

Heroes vs. Villains			
Name	ID	OD	C	CON
Sandra	3	12	0.679	53
Parvati	8	12	0.679	53
Russell	5	12	0.704	57
Jerri	9	12	0.679	53
Colby	7	11	0.655	46
Rupert	10	10	0.679	50
Danielle	4	9	0.633	40
Candice	5	8	0.559	45
Amanda	10	7	0.559	37
J.T.	5	6	0.528	33
Courtney	9	5	0.487	27
Coach	4	4	0.442	19
Rob	5	3	0.432	10
James	7	5	0.452	23
Tyson	3	2	0.422	14
Tom	8	4	0.413	24
Cirie	3	3	0.311	18
Randy	9	1	0.306	4
Stephenie	6	2	0.373	17
Sugar	9	1	0.365	9

KR			
Name	ID	OD	C	CON
Michele	2	6	0.625	25
Aubry	8	9	0.652	35
Tai	4	9	0.625	26
Cydney	5	10	0.682	38
Joe	1	9	0.6	25
Jason	5	9	0.652	20
Julia	7	5	0.5	16
Scot	7	9	0.517	25
Debbie	5	4	0.405	12
Nick	6	1	0.294	4
Neal	0	2	0.348	7
Peter	7	3	0.455	14
Anna	5	1	0.319	5
Alecia	8	3	0.375	11
Caleb	0	0	0	0
Liz	5	1	0.417	5
Jennifer	3	3	0.283	8
Darnell	7	1	0.278	4

Marquesas			
Name	ID	OD	C	CON
Vecepia	2	11	0.778	50
Neleh	4	10	0.737	45
Kathy	5	9	0.7	37
Paschal	0	9	0.714	45
Sean	7	10	0.737	43
Robert	6	6	0.56	25
Tammy	5	5	0.56	25
Zoe	8	4	0.452	24
John	8	3	0.5	18
Rob	8	5	0.438	23
Gina	3	5	0.5	14
Gabriel	7	1	0.311	7
Sarah	11	4	0.519	14
Hunter	4	3	0.359	11
Patricia	5	2	0.359	11
Peter	5	1	0.269	4

Micronesia			
Name	ID	OD	C	CON
Parvati	4	8	0.654	41
Amanda	4	13	0.773	58
Cirie	3	12	0.773	65
Natalie	3	8	0.654	37
Erik	7	11	0.68	45
Alexis	2	6	0.607	32
James	3	5	0.472	28
Jason	8	5	0.486	17
Ozzy	9	8	0.586	47
Eliza	8	3	0.425	20
Ami	4	6	0.567	35
Tracy	7	5	0.515	28
Kathy	0	2	0.286	10
Chet	12	4	0.459	21
Jonathan	0	2	0.419	11
Joel	6	3	0.37	19
Mikey B.	6	2	0.321	8
Yau-Man	6	2	0.415	11
Mary	6	1	0.347	6
Jon	9	1	0.378	7

Millennials vs. Gen X			
Name	ID	OD	C	CON
Adam	6	13	0.655	52
Hannah	13	12	0.655	51
Ken	3	15	0.76	69
David	10	15	0.76	67
Bret	5	14	0.655	51
Jay	10	11	0.633	36
Sunday	5	12	0.655	52
Will	6	9	0.633	41
Zeke	14	7	0.576	37
Jessica	9	9	0.559	44
Chris	7	7	0.514	32
Taylor	7	4	0.487	12
Michelle	10	3	0.463	18
Michaela	4	2	0.432	9
Figgy	6	2	0.463	8
CeCe	11	4	0.5	21
Lucy	2	3	0.432	18
Paul	6	2	0.373	12
Mari	7	1	0.322	4
Rachel	5	1	0.413	4

Nicaragua			
Name	ID	OD	C	CON
Fabio	2	11	0.621	49
Chase	1	10	0.692	66
Sash	2	11	0.692	61
Holly	4	12	0.72	74
Dan	9	11	0.692	58
Jane	11	11	0.621	47
Benry	5	7	0.6	45
Kelly S.	0	7	0.5	34
NaOnka	3	6	0.5	46
Brenda	13	7	0.462	32
Marty	15	7	0.545	46
Alina	10	4	0.474	31
Jill	3	6	0.529	41
Yve	7	5	0.5	28
Kelly B.	8	2	0.383	15
Tyrone	6	4	0.419	21
Jimmy T.	5	3	0.439	22
Jimmy J.	8	2	0.429	15
Shannon	7	1	0.321	11
Wendy	9	1	0.34	6

One World			
Name	ID	OD	C	CON
Kim	3	10	0.667	49
Sabrina	2	10	0.667	50
Chelsea	4	10	0.667	49
Christina	9	11	0.593	32
Alicia	5	10	0.64	38
Tarzan	12	10	0.696	47
Kat	7	7	0.571	29
Troyzan	6	7	0.615	35
Leif	5	7	0.593	28
Jay	5	5	0.552	32
Michael	9	4	0.485	21
Jonas	10	4	0.432	23
Colton	1	3	0.4	16
Monica	5	2	0.457	11
Bill	8	2	0.4	10
Matt	7	1	0.291	0
Nina	6	1	0.372	6
Kourtney	0	0	0	0

Palau			
Name	ID	OD	C	CON
Tom	0	8	0.607	30
Katie	1	7	0.615	29
Ian	5	6	0.615	29
Jenn	3	6	0.593	27
Caryn	7	5	0.593	25
Gregg	4	4	0.552	25
Stephenie	8	11	0.667	40
Janu	1	2	0.485	14
Coby	7	2	0.372	7
Bobby Jon	2	8	0.372	19
Ibrehem	4	8	0.457	23
James	7	7	0.372	20
Angie	8	5	0.308	19
Willard	8	1	0.39	0
Kim	8	4	0.296	18
Jeff	5	3	0.291	14
Ashlee	6	2	0.239	9
Jolanda	6	1	0.239	4
Wanda	0	0	0	0
Jonathan	0	0	0	0

Panama			
Name	ID	OD	C	CON
Aras	9	8	0.682	35
Danielle	4	9	0.714	36
Terry	1	9	0.625	23
Cirie	3	9	0.682	34
Shane	9	7	0.625	28
Courtney	4	6	0.577	28
Bruce	2	4	0.469	21
Sally	8	5	0.484	15
Austin	7	5	0.536	21
Nick	6	4	0.484	17
Dan	3	3	0.385	17
Bobby	3	2	0.417	5
Ruth-Marie	6	3	0.395	13
Misty	5	1	0.288	4
Melinda	5	2	0.429	9
Tina	3	1	0.417	2

Pearl Islands			
Name	ID	OD	C	CON
Sandra	0	11	0.75	53
Lillian	10	11	0.778	50
Jon	7	11	0.778	54
Darrah	6	10	0.7	49
Burton	10	7	0.636	34
Christa	7	9	0.737	47
Tijuana	5	7	0.667	44
Rupert	7	6	0.583	28
Ryan O.	8	5	0.609	30
Andrew	6	4	0.56	25
Osten	2	3	0.5	19
Shawn	5	4	0.519	21
Trish	4	3	0.483	17
Michelle	6	2	0.467	9
Ryan S.	6	2	0.35	7
Nicole	7	1	0.264	5

Philippines			
Name	ID	OD	C	CON
Denise	6	14	0.875	50
Lisa	0	8	0.6	25
Michael	0	8	0.625	29
Malcolm	6	12	0.778	42
Abi-Maria	9	7	0.609	24
Carter	5	8	0.667	28
Jonathan	14	7	0.636	25
Pete	9	4	0.583	19
Artis	5	3	0.452	14
Jeff	5	4	0.452	15
R.C.	4	1	0.378	5
Katie	4	2	0.412	11
Dawson	5	1	0.483	4
Dana	0	0	0	0
Russell	4	4	0.5	15
Angie	4	3	0.359	10
Roxanne	4	2	0.275	7
Zane	5	1	0.341	3

Redemption Island			
Name	ID	OD	C	CON
Rob	7	13	0.739	66
Phillip	17	13	0.739	58
Natalie	1	13	0.708	55
Ashley	3	13	0.708	64
Andrea	9	11	0.654	56
Grant	10	11	0.708	56
Steve	9	11	0.63	41
Ralph	10	9	0.63	44
Julie	6	9	0.63	41
David	8	8	0.607	37
Mike	6	7	0.586	37
Matthew	10	3	0.531	22
Sarita	6	5	0.447	21
Stephanie	8	3	0.436	20
Krista	6	3	0.425	16
Kristina	9	3	0.436	11
Russell	8	1	0.395	8
Francesca	4	1	0.436	11

Samoa			
Name	ID	OD	C	CON
Natalie	8	14	0.85	76
Russell H.	9	15	0.85	76
Mick	4	15	0.81	70
Brett	3	11	0.68	52
Jaison	7	14	0.81	71
Shambo	6	9	0.607	31
Monica	7	8	0.607	38
Dave	8	7	0.607	37
John	7	6	0.607	34
Laura	10	4	0.548	26
Kelly	4	3	0.515	22
Erik	10	2	0.486	13
Liz	5	5	0.515	31
Russell S.	0	1	0.29	7
Ashley	9	4	0.37	22
Yasmin	8	1	0.386	5
Ben	7	3	0.283	19
Betsy	7	2	0.279	13
Mike	0	1	0.225	6
Marisa	7	1	0.274	7

San Juan del Sur			
Name	ID	OD	C	CON
Natalie	0	10	0.63	40
Jaclyn	6	14	0.696	44
Missy	3	12	0.667	34
Keith	16	11	0.667	34
Baylor	17	13	0.762	48
Jon	8	9	0.593	27
Alec	4	9	0.615	41
Reed	10	5	0.593	22
Wes	2	7	0.593	35
Jeremy	5	3	0.516	15
Josh	6	5	0.485	19
Julie	2	1	0.262	4
Dale	8	6	0.533	21
Kelley	4	2	0.471	14
Drew	5	1	0.327	3
John	5	4	0.485	19
Val	9	2	0.485	14
Nadiya	5	1	0.356	4

South Pacific			
Name	ID	OD	C	CON
Sophie	5	14	0.727	74
Coach	0	14	0.773	80
Albert	1	14	0.762	78
Ozzy	17	9	0.615	34
Rick	15	12	0.762	72
Brandon	4	12	0.762	72
Edna	15	11	0.696	58
Cochran	13	11	0.667	60
Whitney	7	10	0.615	58
Dawn	9	9	0.593	43
Jim	8	8	0.593	47
Keith	12	5	0.571	39
Mikayla	4	3	0.471	19
Elyse	3	3	0.432	21
Stacey	10	2	0.457	14
Mark	6	2	0.4	14
Christine	4	1	0.432	10
Semhar	8	1	0.41	9

Thailand			
Name	ID	OD	C	CON
Brian	0	9	0.714	40
Clay	3	8	0.667	33
Jan	5	8	0.7	35
Helen	5	8	0.7	33
Ted	7	7	0.667	30
Jake	7	8	0.667	29
Penny	5	7	0.609	31
Ken	6	6	0.583	26
Erin	3	5	0.438	23
Shii Ann	10	4	0.424	16
Robb	5	3	0.304	6
Stephanie	5	2	0.304	6
Ghandia	5	3	0.467	11
Jed	5	1	0.304	6
Tanya	5	2	0.452	9
John	6	1	0.326	4

The Amazon			
Name	ID	OD	C	CON
Jenna	3	11	0.75	50
Matthew	6	10	0.714	48
Rob	4	10	0.75	51
Butch	5	10	0.75	41
Heidi	3	9	0.714	45
Christy	9	8	0.625	27
Alex	6	7	0.577	32
Deena	6	6	0.517	31
Dave	8	5	0.536	21
Roger	11	4	0.5	20
Shawna	6	3	0.517	16
Jeanne	5	3	0.517	13
JoAnna	4	2	0.441	13
Daniel	7	2	0.341	8
Janet	5	1	0.349	4
Ryan	4	1	0.341	8

The Australian Outback			
Name	ID	OD	C	CON
Tina	0	12	0.778	43
Colby	10	12	0.813	46
Keith	10	11	0.765	39
Elisabeth	5	10	0.65	32
Rodger	5	10	0.65	32
Amber	6	10	0.722	36
Nick	4	7	0.619	29
Jerri	12	8	0.65	30
Alicia	5	5	0.565	23
Jeff	11	3	0.542	15
Michael	0	2	0.292	11
Kimmi	6	2	0.371	12
Mitchell	6	3	0.5	16
Maralyn	5	2	0.419	14
Kel	7	1	0.406	8
Debb	7	1	0.361	6

Tocantins			
Name	ID	OD	C	CON
J.T.	0	11	0.714	41
Stephen	1	10	0.778	46
Erinn	5	8	0.7	36
Taj	7	10	0.737	32
Coach	6	7	0.609	29
Debbie	6	6	0.467	25
Sierra	11	5	0.467	24
Tyson	5	4	0.368	17
Brendan	4	3	0.412	16
Joe	1	4	0.5	21
Sydney	4	4	0.5	21
Spencer	5	3	0.483	17
Sandy	6	2	0.35	9
Jerry	6	2	0.438	9
Candace	7	1	0.326	6
Carolina	7	1	0.264	5

Vanuatu			
Name	ID	OD	C	CON
Chris	3	12	0.708	48
Twila	6	10	0.708	57
Scout	3	11	0.68	58
Eliza	9	11	0.654	49
Julie	4	9	0.654	44
Ami	8	9	0.654	47
Leann	8	8	0.607	45
Chad	7	7	0.63	33
Lea	7	6	0.586	30
Rory	14	6	0.548	27
John K.	5	4	0.515	23
Lisa	4	4	0.5	26
Travis	6	4	0.415	23
Brady	6	3	0.362	10
Mia	5	2	0.459	9
John P.	5	2	0.472	12
Dolly	5	1	0.386	7
Brook	5	1	0.425	2

Worlds Apart			
Name	ID	OD	C	CON
Mike	4	12	0.68	45
Carolyn	10	11	0.68	44
Will	4	12	0.68	45
Rodney	5	11	0.63	39
Sierra	6	10	0.63	41
Dan	9	11	0.654	40
Tyler	5	8	0.607	36
Shirin	7	7	0.607	36
Jenn	18	7	0.607	24
Joe	8	6	0.531	26
Hali	8	5	0.548	19
Kelly	4	4	0.548	21
Joaquin	4	2	0.472	12
Max	5	2	0.436	6
Lindsey	5	1	0.405	3
Nina	4	2	0.436	14
Vince	3	1	0.386	11
So	4	1	0.415	6

Big Brother 1 (US)			
Name	ID	OD	C	CON
Eddie	17	14	0.75	28
Josh	10	14	0.75	25
Curtis	18	14	0.692	22
Jamie	7	14	0.9	37
George	9	12	0.818	30
Cassandra	8	10	0.75	24
Brittany	6	8	0.75	29
Karen	7	6	0.692	24
Jordan	10	4	0.529	8
William	6	2	0.529	7

Big Brother 2 (US)			
Name	ID	OD	C	CON
Will	1	5	0.625	10
Nicole	3	6	0.667	18
Monica	1	6	0.714	19
Hardy	1	5	0.476	16
Bunky	2	6	0.5	23
Krista	4	3	0.4	12
Kent	4	2	0.357	9
Mike	4	2	0.357	11
Shannon	6	2	0.455	8
Autumn	7	1	1.0	4
Sheryl	5	0	0.0	0

Big Brother 3 (US)			
Name	ID	OD	C	CON
Lisa	3	8	0.714	19
Danielle	0	9	0.688	24
Jason	1	9	0.769	24
Amy	13	1	0.667	4
Marcellas	2	5	0.588	18
Roddy	4	4	0.588	13
Gerry	4	6	0.5	17
Chiara	4	4	0.556	17
Josh	8	3	0.476	8
Eric	4	3	1.0	14
Tonya	5	1	1.0	4
Lori	5	0	0.0	0

Big Brother 4 (US)			
Name	ID	OD	C	CON
Jun	0	6	0.588	25
Alison	0	9	0.833	25
Robert	2	6	0.643	23
Erika	4	5	0.692	20
Jee	2	4	0.529	14
Jack	4	4	0.5	20
Justin	3	4	0.6	19
Nathan	5	3	0.8	14
Dana	6	2	1.0	13
David	5	2	1.0	13
Michelle	6	1	1.0	8
Amanda	9	0	0.0	0

Big Brother 5 (US)			
Name	ID	OD	C	CON
Drew	0	9	0.8	37
Michael	1	7	0.625	29
Diane	1	9	0.733	34
Nakomis	0	5	0.611	24
Karen	4	8	0.769	35
Marvin	10	3	0.455	18
Adria	5	6	0.667	26
Natalie	4	2	0.526	8
Will	4	5	0.625	25
Jase	6	3	0.556	15
Scott	4	3	1.0	21
Holly	7	2	1.0	15
Lori	7	1	1.0	9
Mike	10	0	0.0	0

Big Brother 6 (US)			
Name	ID	OD	C	CON
Maggie	4	9	0.722	43
Ivette	2	10	0.765	39
Janelle	3	6	0.619	26
April	2	9	0.722	39
Howie	2	6	0.481	33
Beau	2	8	0.65	40
James	4	8	0.65	32
Rachel	5	6	0.481	30
Jennifer	5	5	0.565	33
Kaysar	16	1	0.394	2
Sarah	6	4	0.361	29
Eric	5	1	1.0	8
Michael	9	1	1.0	8
Ashlea	9	0	0.0	0

Big Brother 8 (US)			
Name	ID	OD	C	CON
Dick	3	8	0.75	27
Daniele	0	8	0.684	38
Zach	2	8	0.632	32
Jameka	2	7	0.706	36
Eric	5	8	0.667	33
Jessica	2	6	0.6	37
Amber	4	6	0.667	30
Jen	5	5	0.571	20
Dustin	4	4	0.444	28
Kail	8	1	0.462	2
Nick	6	3	1.0	23
Mike	7	2	1.0	17
Joe	9	1	1.0	9
Carol	10	0	0.0	0

Big Brother 9 (US)			
Name	ID	OD	C	CON
Adam	0	10	0.737	41
Ryan	9	7	0.65	22
Sheila	1	11	0.813	42
Sharon	6	5	0.52	23
Natalie	2	9	0.565	19
James	8	8	0.684	31
Joshuah	3	7	0.619	32
Chelsia	5	7	0.65	27
Matt	4	6	0.542	15
Allison	12	2	1.0	10
Alex	6	0	0.0	0
Amanda	6	0	0.0	0
Jen	6	2	0.371	5
Parker	6	2	0.371	5
Jacob	2	0	0.0	0

Big Brother 10 (US)			
Name	ID	OD	C	CON
Dan	0	8	0.667	22
Memphis	3	9	0.846	36
Jerry	3	4	0.611	22
Keesha	1	5	0.524	24
Renny	3	6	0.647	27
Ollie	3	7	0.688	31
Michelle	3	5	0.579	28
April	4	4	0.833	28
Libra	6	4	1.0	26
Jessie	4	1	1.0	8
Angie	8	2	1.0	16
Steven	9	1	1.0	8
Brian	9	0	0.0	0

Big Brother 11 (US)			
Name	ID	OD	C	CON
Jordan	2	5	0.632	11
Natalie	3	5	0.6	19
Kevin	1	9	0.8	29
Michele	1	8	0.706	31
Jeff	2	5	0.414	26
Russell	3	5	0.5	26
Lydia	6	4	0.462	21
Chima	5	3	0.353	19
Jessie	3	3	0.375	17
Ronnie	4	2	0.25	12
Casey	8	2	0.273	11
Laura	8	1	0.267	4
Braden	6	0	0.0	0

Big Brother 12 (US)			
Name	ID	OD	C	CON
Hayden	1	6	0.688	27
Lane	0	7	0.462	39
Enzo	1	9	0.846	39
Britney	1	7	0.688	36
Ragan	2	8	0.647	43
Brendon	3	5	0.407	32
Matt	6	3	0.423	18
Kathy	5	4	0.524	19
Rachel	6	1	0.6	7
Kristen	6	3	1.0	22
Andrew	8	2	1.0	15
Monet	7	1	1.0	9
Annie	10	0	0.0	0

Big Brother 13 (US)			
Name	ID	OD	C	CON
Rachel	2	7	0.667	17
Porsche	5	8	0.706	31
Adam	3	8	0.706	29
Jordan	3	8	0.75	30
Kalia	4	7	0.6	29
Shelly	3	6	0.571	27
Jeff	3	6	0.414	30
Daniele	3	4	0.522	20
Brendon	10	3	0.364	19
Lawon	6	4	0.5	24
Dominic	7	2	0.444	12
Cassi	9	1	0.429	4
Keith	6	0	0.0	0

Big Brother 14 (US)			
Name	ID	OD	C	CON
Ian	0	8	0.688	29
Dan	0	9	0.765	30
Danielle	3	5	0.588	17
Shane	1	9	0.786	31
Jenn	4	8	0.75	26
Joe	7	5	0.692	19
Frank	7	2	0.529	7
Britney	4	4	0.6	20
Ashley	5	5	0.643	22
Mike	5	2	0.333	12
Wil	6	3	0.391	15
Janelle	8	0	0.0	0
JoJo	5	1	0.375	4
Kara	5	0	0.0	0
Jodi	1	0	0.0	0

Big Brother 15 (US)			
Name	ID	OD	C	CON
Andy	0	12	0.867	63
GinaMarie	0	10	0.737	42
Spencer	5	6	0.571	39
McCrae	1	11	0.813	56
Judd	9	8	0.75	47
Elissa	12	8	0.667	43
Amanda	3	8	0.75	46
Aaryn	5	3	0.48	20
Helen	4	6	0.667	38
Jessie	6	6	0.632	41
Candice	8	4	0.5	30
Howard	7	4	0.48	30
Kaitlin	9	3	0.48	11
Jeremy	9	2	0.414	8
Nick	7	1	1.0	6
David	7	0	0.0	0

Big Brother 16 (US)			
Name	ID	OD	C	CON
Derrick	0	11	0.857	65
Cody	0	12	0.778	54
Victoria	1	10	1.0	74
Caleb	1	8	0.786	50
Frankie	2	10	0.846	62
Christine	3	10	1.0	74
Nicole	12	5	1.0	48
Donny	5	7	0.727	43
Zach	9	6	0.875	50
Hayden	5	6	0.875	54
Jocasta	6	4	0.667	37
Amber	9	4	1.0	40
Brittany	10	3	1.0	31
Devin	11	1	1.0	12
Paola	10	0	0.0	0
Joey	13	0	0.0	0

Big Brother 17 (US)			
Name	ID	OD	C	CON
Steve	1	11	0.75	63
Liz	0	11	0.762	69
Vanessa	1	11	0.778	59
John	7	10	0.813	60
Austin	2	12	0.929	75
Julia	4	6	0.65	37
James	8	8	0.684	40
Meg	7	8	0.684	50
Becky	8	5	0.542	39
Jackie	7	5	0.591	35
Shelli	8	3	0.371	23
Clay	9	5	0.448	37
Jason	7	4	0.5	31
Audrey	9	3	0.5	18
Jeff	7	1	1.0	11
Da'Vonne	7	1	1.0	11
Jace	12	0	0.0	0

Big Brother 18 (US)			
Name	ID	OD	C	CON
Nicole	0	12	0.667	53
Paul	2	9	0.722	37
James	1	12	0.765	44
Corey	2	8	0.684	47
Victor	13	5	0.692	29
Natalie	3	9	0.722	43
Michelle	6	7	0.818	48
Paulie	9	5	0.692	32
Bridgette	7	3	0.6	19
Zakiyah	3	6	0.75	42
Da'Vonne	6	4	0.6	27
Frank	9	4	0.529	23
Tiffany	12	1	1.0	6
Bronte	6	1	0.45	6
Jozea	7	0	0.0	0

Big Brother 19 (US)			
Name	ID	OD	C	CON
Josh	3	12	0.789	61
Paul	11	12	0.789	60
Christmas	6	12	0.789	61
Kevin	5	11	0.714	70
Alex	3	9	0.682	48
Raven	3	10	0.714	60
Jason	3	9	0.652	51
Matt	8	7	0.6	49
Mark	4	9	0.652	50
Elena	6	6	0.625	45
Cody	14	3	0.536	12
Jessica	7	4	0.556	28
Ramses	10	4	0.556	30
Dominique	10	3	0.395	24
Jillian	11	0	0.0	0
Megan	0	1	1.0	7
Cameron	8	0	0.0	0

Big Brother All-Stars (US)			
Name	ID	OD	C	CON
Mike	0	9	0.765	45
Erika	3	6	0.667	31
Janelle	1	7	0.667	25
Will	1	9	0.846	37
George	3	6	0.5	39
Danielle	5	6	0.667	32
James	5	5	0.625	29
Howie	3	6	0.667	38
Marcellas	6	5	0.455	34
Kaysar	5	3	0.4	23
Diane	9	2	0.435	12
Jase	9	1	0.313	7
Nakomis	8	1	0.417	4
Alison	8	0	0.0	0

Big Brother Over The Top			
Name	ID	OD	C	CON
Morgan	0	6	0.667	21
Jason	0	7	0.733	24
Kryssie	1	4	0.571	11
Justin	2	7	0.75	24
Shelby	2	5	0.692	21
Danielle	11	3	0.583	8
Whitney	3	4	0.636	22
Alex	3	4	0.7	17
Scott	5	3	0.583	18
Neeley	3	2	1.0	10
Shane	5	2	1.0	10
Monte	4	0	0.0	0
Cornbread	8	0	0.0	0

A.2   Complete Alliances Data

Season	Winner	Finalists	Alliances	ED	Full ED
Africa	Ethan	Kim J.	*Older Samburu*: Frank, Teresa, Linda, Carl	0.167	0.85
			Boran: Lex, Ethan, Kim J., Tom, Kelly	1.0	
			Younger Samburu: Silas, Kim P., Brandon, Lindsey	0.167	
All-Stars	Amber	Rob M.	*Chapera*: Rob M., Amber, Jenna L., Rupert, Tom, Alicia	1.067	0.581
			Mogo Mogo: Lex, Shii Ann, Kathy, Jerri	0.667	
Blood vs. Water	Tyson	Monica, Gervase	*Singles*: Tyson, Monica, Gervase, Ciera, Hayden, Caleb	1.667	0.747
			Five Guys: Hayden, Brad, John, Caleb, Vytas	0.9	
			Galang: Tina, Aras, Tyson, Monica, Gervase	1.3	
Borneo	Richard	Kelly	*Barbecue*: Colleen, Jenna, Gervase	1.667	0.783
			Tagi: Richard, Rudy, Susan, Kelly	1.5	
Cagayan	Tony	Woo	*Solana*: Trish, Jefra, LJ, Tony, Woo	0.7	0.647
			Aparri: Spencer, Tasha, Jeremiah, Morgan, Sarah, Kass	0.533	
Cambodia	Jeremy	Spencer, Tasha	*Bayon*: Jeremy, Tasha, Stephen, Andrew, Keith, Joe, Kimmi	0.905	0.705
			Witches' Coven: Kelley, Abi-Maria, Ciera, Kass	0.167	
Caramoan	Cochran	Dawn, Sherri	*Stealth R Us*: Cochran, Dawn, Phillip, Andrea, Malcolm, Corinne	0.8	0.742
			Gota: Sherri, Julia, Shamar, Laura, Michael, Matt	0.667	
			Cool Kids: Eddie, Reynold, Hope, Allie	0.0	
China	Todd	Courtney, Amanda	*Fei Long*: Todd, Courtney, Amanda, Aaron, Denise, James, Frosti	0.667	0.75
			Zhan Hu: Peih-Gee, Erik, Jaime	0.0	
Cook Islands	Yul	Ozzy, Becky	*Aitu Four*: Yul, Ozzy, Becky, Sundra	1.0	0.611
			Raro: Adam, Parvati, Candice, Nate, Jonathan	0.9	
Fiji	Earl	Cassandra, Dreamz	*Four Horsemen*: Alex, Mookie, Edgardo, Dreamz	0.667	0.66
			Syndicate: Earl, Cassandra, Michelle, Yau-Man	0.667	

Season	Winner	Finalists	Alliances	ED	Full ED
Gabon	Bob	Susie, Sugar	*Onion*: Bob, Corinne, Randy, Charlie, Marcus, Jacquie, Susie	0.476	0.667
			Fang: Ken, Crystal, Kelly, G.C., Susie, Sugar, Matty	1.095	
Game Changers	Sarah	Brad, Troyzan	*Power Six*: Sarah, Brad, Troyzan, Sierra, Debbie, Tai	0.933	0.737
			Tavua: Aubry, Cirie, Michaela, Ozzy, Andrea, Zeke, Sarah	1.238	
Guatemala	Danni	Stephenie	*Nakum*: Stephenie, Rafe, Lydia, Cindy, Judd, Jamie	1.067	0.712
			Yaxha: Danni, Bobby Jon, Brandon, Blake	0.333	
Heroes vs. Healers vs. Hustlers	Ben	Chrissy, Ryan	*Healers*: Joe, Desi, Jessica, Cole, Mike	0.6	0.706
			The Round Table: Chrissy, Ryan, Devon, JP, Ben, Ashley, Lauren	0.905	
			Final Four: Ashley, Lauren, Ben, Devon	1.333	
Heroes vs. Villains	Sandra	Parvati, Russell	*Heroes*: Rupert, Amanda, J.T., Cirie, James, Candice	0.533	0.679
			Rob's Villains: Sandra, Courtney, Rob, Tyson, Jerri, Coach	0.667	
			Russell's Villains: Parvati, Russell, Jerri, Danielle	1.0	
KR	Michele	Aubry, Tai	*Gondol*: Jason, Julia, Scot, Tai	1.0	0.625
			Dara Women: Michele, Aubry, Joe, Cydney, Julia, Debbie	0.8	
Marquesas	Vecepia	Neleh	*Maraamu*: Rob, Vecepia, Sean, Rob, Sarah	0.3	0.733
			Rotu Four: John, Robert, Tammy, Zoe	0.667	
			Outsiders: Kathy, Vecepia, Neleh, Paschal, Sean	1.0	
Micronesia	Parvati	Amanda	*Black Widow Brigade*: Parvati, Amanda, Cirie, Natalie, Alexis	1.0	0.563
			Malakal Couples: James, Ozzy, Parvati, Amanda, Cirie	0.5	
			Older Airai: Tracy, Kathy, Chet	0.0	
Millennials vs. Gen X	Adam	Hannah, Ken	*Triforce*: Jay, Will, Taylor, Michelle, Figgy, Michaela	0.267	0.768
			David's Vinaka: Hannah, Ken, Jessica, David, Adam	0.6	
			Zeke's Vinaka: Bret, Jay, Sunday, Zeke, Will	0.7	
			Takali: Bret, Sunday, Chris, Lucy, Paul, Jessica	0.733	
Nicaragua	Fabio	Chase, Sash	*Final Four*: Chase, Sash, NaOnka, Holly, Jane	0.7	0.674
			La Flor: Kelly S., Brenda, Chase, Sash, NaOnka	0.4	
			Espada: Dan, Marty, Jill	0.0	
One World	Kim	Sabrina, Chelsea	*Misfit*: Leif, Jonas, Colton, Tarzan, Troyzan	0.3	0.765
			Muscle: Michael, Matt, Jay, Bill	1.0	
			Salani: Kim, Sabrina, Chelsea, Alicia, Kat	0.9	
Palau	Tom	Katie	*Koror*: Tom, Katie, Ian, Jenn, Gregg	1.0	0.588

Season	Winner	Finalists	Alliances	ED	Full ED
Panama	Aras	Danielle	*La Mina*: Terry, Austin, Nick, Dan	0.667	0.65
			Casaya: Shane, Courtney, Bruce, Aras, Danielle, Cirie	1.067	
Pearl Islands	Sandra	Lillian	*Morgan*: Andrew, Ryan O., Osten, Darrah, Tijuana	0.2	0.8
			Drake: Rupert, Sandra, Christa, Jon, Trish	0.8	
			Outcast: Burton, Jon, Tijuana, Lillian, Darrah	1.9	
Philippines	Denise	Lisa, Michael	*Matsing*: Denise, Malcolm, Angie	1.333	0.654
			Kalabaw: Carter, Jonathan, Jeff	0.0	
			Fulcrum: Michael, Lisa, R.C	0.0	
			Tandang: Abi-Maria, Pete, Artis, R.C., Lisa, Michael	0.533	
Redemption Island	Rob	Philip, Natalie	*Zapatera Six*: Mike, Ralph, Steve, Julie, David, Sarita	0.467	0.895
			Stealth R Us: Rob, Phillip, Natalie, Ashley, Andrea, Matthew, Grant	1.571	
			Russell's Zapatera: Stephanie, Krista, Russell	0.0	
Samoa	Natalie	Russell H., Mick	*Galu*: Brett, Monica, Dave, Laura, Kelly, Shambo, John	0.667	0.663
			Foa Foa Four: Natalie, Russell H., Mick, Jaison	0.667	
San Juan del Sur	Natalie	Jaclyn, Missy	*Fab Five*: Missy, Baylor, Jaclyn, Jon, Natalie	1.2	0.752
			Coyopa Guys: Alec, Wes, Josh, Dale, John	0.3	
South Pacific	Sophie	Coach, Albert	*The Family*: Sophie, Coach, Albert, Rick, Brandon, Edna	1.2	0.922
			Savaii: Ozzy, Whitney, Dawn, Keith, Jim, Elyse, Cochran	1.143	
Thailand	Brian	Clay	*Sook Jai*: Jake, Penny, Ken, Erin, Shii Ann	1.0	0.683
			Chuay Gahn Five: Brian, Clay, Jan, Helen, Ted	1.2	
The Amazon	Jenna	Matthew	*Tambaqui*: Roger, Matthew, Rob, Butch, Dave, Alex	1.133	0.767
			Jaburu: Jenna, Heidi, Alex, Deena, Shawna, Rob, Matthew	1.048	
The Australian Outback	Tina	Colby	*Ogakor*: Amber, Jerri, Tina, Colby, Keith	1.0	0.825
			Kucha: Elisabeth, Rodger, Nick, Alicia, Jeff	0.0	
Tocantins	J.T.	Stephen	*Exile*: Sierra, Brendan, Stephen, Taj	0.667	0.675
			Timbira: Coach, Tyson, Debbie	0.333	
			Jalapao Three: J.T., Stephen, Taj	0.667	
Vanuatu	Chris	Twila	*Yasur*: Ami, Leann, Lisa, Twila, Scout, Eliza	1.2	0.719
			Final Four: Chris, Twila, Scout, Eliza	1.333	
			Fat Five: Chris, Chad, Lea, Rory, Travis	0.3	
Worlds Apart	Mike	Carolyn, Will	*Escameca*: Rodney, Dan, Kelly, Mike, Sierra	1.0	0.739
			Nagarote: Jenn, Joe, Hali, Will	1.0	

Season	Winner	Finalists	Alliances	ED	Full ED
10	Dan	Memphis	*The Renegades*: Dan, Memphis	0.0	0.718
			The Coven: Keesha, Libra, April	0.667	
12	Hayden	Lane	*The Brigade*: Enzo, Hayden, Lane, Matt	0.5	0.718
13	Rachel	Porsche	*The Regulators*: Dominic, Keith, Cassi, Lawon	0.5	0.821
			Newbies: Adam, Porsche, Kalia, Shelly, Lawon, Dominic, Cassi, Keith	0.786	
			Veterans: Daniele, Dick, Jordan, Rachel, Jeff, Brendon	0.467	
14	Ian	Dan	*Silent Six*: Britney, Danielle, Dan, Frank, Mike, Shane	0.8	0.581
			Team Toche: Britney, Shane, Danielle, Dan	0.667	
			The Quack Pack: Britney, Danielle, Dan, Ian, Shane	0.6	
			Chilltown 2.0: Mike, Frank	0.0	
15	Andy	GinaMarie	*Tenexas*: Judd, Jessie	1.0	0.767
			Exterminators: GinaMarie, Andy, Judd, Spencer	0.667	
			Young Grasshoppers: GinaMarie, Howard, Andy, Kaitlin, Spencer, Judd	0.733	
			The Moving Company: Nick, McCrae, Spencer, Jeremy, Howard	0.6	
			3 A.M.: Aaryn, Amanda, Andy, McCrae	0.5	
			The Blonde-Tourage: David, Aaryn, Kaitlin, Jeremy, GinaMarie, Jessie	0.4	
			The Goof Troupe: Amanda, Andy, McCrae, Judd	1.167	
			The Mom Squad: Elissa, Helen	0.0	

Season	Winner	Finalists	Alliances	ED	Full ED
16	Derrick	Cody	*The Detonators*: Christine, Cody, Derrick, Frankie, Zach	0.6	0.808
			The Crazy 8's: Amber, Cody, Devin, Donny, Frankie, Joey, Nicole, Paola	0.929	
			El Cuatro: Amber, Joey, Nicole, Paola	0.833	
			Team America: Derrick, Donny, Frankie, Joey	1.0	
			The Bomb Squad: Amber, Caleb, Christine, Cody, Derrick, Devin, Frankie, Hayden, Zach	0.722	
			Los Tres Amigos: Cody, Derrick, Zach	0.333	
			The Hitmen: Cody, Derrick	0.0	
			Zankie: Frankie, Zach	0.0	
			The Rationale: Cody, Derrick, Hayden, Nicole	1.0	
			The Double D's: Donny, Devin	1.0	
			The Weirdos: Christine, Hayden, Nicole	0.667	
17	Steve	Liz	*Clelli*: Clay, Shelli	0.0	0.765
			ShellTown: Jace, Austin	1.0	
			Team JJ: Jackie, Jeff	0.0	
			Students of Sound: Steve, Vanessa	1.0	
			Three's Company: Clay, Shelli, Vanessa	0.333	
			The Goblins: James, Jackie, Meg, Audrey, DaVonne, Jason, Jeff	0.667	
			The Sixth Sense: Clay, Austin, Julia, Liz, Shelli, Vanessa	0.667	
			Jecky: Becky, John	0.0	
			Austwins: Austin, Julia, Liz, Jace	0.5	
			Scamper Squad: Vanessa, Liz, Steve, Austin, Julia	0.5	
			Rockstars: John, Steve	0.0	
18	Nicole	Paul	*The Revolution*: Paul, Jozea, Victor	0.0	0.819
			Team PP: Paulie, Paul	2.0	
			Nicorey: Corey, Nicole	0.0	
			Zaulie: Paulie, Zakiyah	0.0	
			Spy Girls: Natalie, Bronte, Bridgette	0.0	
			Fatal Five: DaVonne, Michelle, Nicole, Tiffany, Zakiyah	0.6	
			Final Four: Corey, Nicole, Paul, Victor	0.833	
			8-Pack: Nicole, James, DaVonne, Frank, Tiffany, Corey, Michelle, Zakiyah	0.75	
			Jatalie: James, Natalie	0.0	
			The Executives: Corey, James, Paul, Paulie, Victor	1.2	
			The Sitting Ducks: Paul, Victor	0.0	
19	Josh	Pau	*The Team*: Christmas, Cody, Dominique, Elena, Jessica, Mark, Matt, Paul, Raven	1.028	0.824
			Marlena: Elena, Mark	0.0	
			Jody: Cody, Jessica	0.0	
			The Misfits: Christmas, Josh, Paul	1.0	
			Whistlenut and Ole: Alex, Jason, Paul	1.0	
2	Will	Nicole	*Chilltown*: Mike, Will, Shannon	0.333	0.691
			TOP (The Other People): Bunky, Kent, Nicole, Hardy, Monica, Autumn	0.733	

Season	Winner	Finalists	Alliances	ED	Full ED
3	Lisa	Danielle	*Danielle and Jason*: Danielle, Jason	0.0	0.803
			Chiara and Roddy: Chiara, Roddy	0.0	
			Cartel: Lisa, Chiara, Tonya, Roddy, Eric, Josh	0.667	
			Original Six: Josh, Roddy, Lisa, Chiara, Eric, Gerry	0.933	
			Eric and Lisa: Eric, Lisa	0.0	
4	Jun	Alison	*Three Stooges*: Jee, Justin, Robert	0.333	0.697
			Girl Power: Alison, Erika, Jun	0.333	
			Elite Eight: Alison, Dana, David, Erika, Jack, Jun, Nathan, Scott	0.464	
5	Drew	Michael	*Four Horsemen*: Drew, Jase, Michael, Scott	0.167	0.692
			Pinky Swear: Adria, Diane, Karen, Natalie, Nakomis, Will	0.667	
6	Ivette	Maggie	*Sovereign Six*: Janelle, Howie, Kaysar, James	1.333	0.813
			The Friendship: Ivette, Maggie, Eric, Beau, Jennifer, April	0.667	
8	Dick	Daniele	*Mrs. Robinson*: Zach, Kail, Nick, Mike	0.5	0.736
			Late Night Crew: Amber, Daniele, Dick, Dustin, Eric, Jameka, Jessica	0.619	
9	Adam	Ryan	*Team Christ*: Adam, Ryan, Sheila, Natalie	1.333	0.724
All-Stars	Mike	Erika	*Sovereign Six*: Janelle, Howie, Kaysar, James	0.5	0.725
			Chilltown: Mike, Will	0.0	
			Mr. and Mrs. Smith: Diane, Jase	1.0	
			The Legion of Doom: Danielle, James, Mike, Will	0.5	
Over The Top	Morgan	Jason, Kryssie	*OTT Jamboree*: Jason, Justin, Kryssie, Scott, Shelby	0.6	0.603
			Boys Alliance: Cornbread, Monte, Scott, Shane	0.667	
			The Ballsmashers: Alex, Morgan, Shelby, Whitney	0.0	
			Late Night Jamboree: Danielle, Jason, Justin, Kryssie, Shane	0.6	
			The Southerners: Alex, Monte, Morgan, Shane, Whitney	0.4	
			The Jackolanterns: Jason, Kryssie, Neeley	0.333	
			Shonte: Monte, Shane	1.0	
			Shanielle: Danielle, Shane	0.0	
			Monte and His Pythons: Alex, Monte, Morgan, Shelby, Whitney	0.0	
			Krason: Jason, Kryssie	1.0	
			Sisters: Alex, Morgan	0.0	
			Team Longshot: Morgan, Shelby	0.0	

References

1. Bastian, M., Heymann, S., Jacomy, M.: Gephi: an open source software for exploring and manipulating networks. In: Proceedings of the International AAAI Conference on Weblogs and Social Media (2009)
2. Big Brother Wiki. http://bigbrother.wikia.com/wiki/Big_Brother_Wiki
3. Boginski, V., Butenko, S., Pardalos, P.M.: On structural properties of the market graph. In: Nagurney, A. (ed.) Innovation in Financial and Economic Networks, pp. 29–45. Edward Elgar Publishers (2003)
4. Bonato, A.: A Course on the Web Graph. American Mathematical Society Graduate Studies Series in Mathematics, Providence, Rhode Island (2008)
5. Bonato, A., Gleich, D.F., Kim, M., Mitsche, D., Pralat, P., Tian, A., Young, S.J.: Dimensionality matching of social networks using motifs and eigenvalues. PLOS ONE **9**(9), e106052 (2014)
6. Bonato, A., Hadi, N., Pralat, P., Wang, C.: Dynamic models of on-line social networks. In: Proceedings of WAW 2009 (2009)
7. Bonato, A., Hadi, N., Horn, P., Pralat, P., Wang, C.: Models of on-line social networks. Internet Math. **6**, 285–313 (2011)
8. Bonato, A., Tian, A.: Complex networks and social networks. In: Kranakis, E. (ed.) Advances in Network Analysis and its Applications. MATHINDUSTRY, vol. 18, pp. 269–286. Springer, Berlin (2012). https://doi.org/10.1007/978-3-642-30904-5_12
9. Brandes, U., Erlebach, T. (eds.): Network Analysis. LNCS, vol. 3418. Springer, Heidelberg (2005). https://doi.org/10.1007/b106453
10. Easley, D., Kleinberg, J.: Networks, Crowds, and Markets Reasoning about a Highly Connected World. Cambridge University Press, Cambridge (2010)
11. Gleich, D.F.: PageRank beyond the Web. SIAM Rev. **57**(3), 321–363 (2015)
12. Gower, J.C., Warrens, M.J.: Similarity, Dissimilarity, and Distance, Measures of. Wiley StatsRef: Statistics Reference Online (2006)
13. Guo, W., Lu, X., Donate, G.M., Johnson, S.: The spatial ecology of war and peace. Preprint (2018)
14. Heider, F.: The Psychology of Interpersonal Relations. Wiley, New York (1958)
15. Langville, A.N., Meyer, C.D.: Who's #1? The Science of Rating and Ranking. Princeton University Press, Princeton (2012)
16. Leskovec, J., Huttenlocher, D., Kleinberg, J.: Predicting positive and negative links in online social networks. In: Proceedings of the 19th International Conference on World Wide Web (WWW 2010) (2010)
17. McPherson, J.M., Ranger-Moore, J.R.: Evolution on a dancing landscape: organizations and networks in dynamic Blau space. Soc. Forces **70**, 19–42 (1991)
18. Survivor Wiki. http://survivor.wikia.com/wiki/Main_Page
19. Tang, J., Chang, S., Aggarwal, C., Liu, H.: Negative link prediction in social media. In: Proceedings of the Eighth ACM International Conference on Web Search and Data Mining (WSDM 2015) (2015)
20. West, D.B.: Introduction to Graph Theory, 2nd edn. Prentice Hall, Upper Saddle River (2001)
21. Yang, S.-H., Smola, A.J., Long, B., Zha, H., Chang, Y.: Friend or frenemy? predicting signed ties in social networks. In: Proceedings of the 35th International ACM SIGIR Conference on Research and Development in Information Retrieval (SIGIR 2012) (2012)

An Experimental Study of the k-MXT Algorithm with Applications to Clustering Geo-Tagged Data

Colin Cooper[(✉)] and Ngoc Vu[(✉)]

Department of Informatics, King's College London, London, UK
{colin.cooper,ngoc.vu}@kcl.ac.uk

Abstract. We consider a *graph fragmentation process* which can be described as follows. Each vertex v selects the k adjacent vertices which have the largest number of common of neighbours. For each selected neighbour u, we retain the edge (v, u) to form a the subgraph graph S of the input graph. The object of interest are the components of S, the k-Max-Triangle-Neighbour (k-MXT) subgraph, and the vertex clusters they produce in the original graph.

We study the application of this process to clustering in the planted partition model, and on the geometric disk graph formed from geo-tagged photographic data downloaded from Flickr.

In the planted partition model, there are ℓ numbers of partitions, or subgraphs, which are connected densely within each partition but sparser between partitions. The objective is to recover these hidden partitions. We study the case of the planted partition model based on the random graph $G_{n,p}$ with additional edge probability q within the partitions. Theoretical and experimental results show that the 2-MXT algorithm can recover the partitions for any $q/p > 0$ constant provided the density of triangles is high enough.

We apply the k-MXT algorithm experimentally to the problem of clustering geographical data, using London as an example. Given a dataset consisting of geographical coordinates extracted from photographs, we construct a disk graph by connecting every point to other points if and only if theirs distance is at most d. Our experimental results show that the k-MXT algorithm is able to produce clusters which are of comparable to popular clustering algorithms such as DBSCAN (see e.g. Fig. 5).

1 Introduction

A graph fragmentation process is a procedure in which every vertex selects one of its neighbours or itself according to some rule. Each selection forms a directed edge and the entirety of such edges is a directed graph S in which every vertex has out-degree 1, and each component of S is unicyclic. Ignoring orientation, S is a subgraph of the original graph and is usually sparser. Typically S usually consists of many smaller components which we call *fragments*.

© Springer International Publishing AG, part of Springer Nature 2018
A. Bonato et al. (Eds.): WAW 2018, LNCS 10836, pp. 145–169, 2018.
https://doi.org/10.1007/978-3-319-92871-5_10

Graph fragmentation can be considered a *mapping* on the vertex set. For example, consider a complete graph K_n with $V = \{1, 2, \ldots, n\}$. Each vertex i *selects* one of its neighbour or itself with equal probability. Since the graph is complete and the queried vertex i is also included, the vertex that i maps to some j such that $1 \leq j \leq n$. A directed edge (i, j) with i being the source and j the target is added to S. This graph model is known as the *random mapping graph* and has been studied extensively by various authors e.g. Bollobas [1], Frieze and Michal [6], etc.

The rule used for graph fragmentation in the 1-MXT algorithm is to choose that neighbour u of v for which the edge (v, u) is contained in the maximum number of triangles among all edges (v, w) incident with v.

Graph clustering is the task of dividing vertices of the graph into sets of vertices, so that there is a higher degree of connectivity for vertices within a cluster, and lower for vertices between different clusters. Inspired by small-world networks in which pairs of adjacent vertices are more likely to be connected to a common neighbour, we consider a fragmentation process in which each vertex selects the k adjacent vertices which share the highest number of common neighbours. The motivation is that the resulting components in the subgraph S should contain many triangles, and thus be dense internally. We call this algorithm k-MaX-Triangles (k-MXT). The algorithm is introduced in Sect. 2.

To test the functionality of the k-MXT algorithm, we use a graph model called the *planted ℓ partition* model. It was first introduced by Condon and Karp in [3] and later considered by Girvan and Newman in [7]. Thereafter it became a standard benchmark graph for testing graph clustering algorithm. A *random planted ℓ partition* graph is constructed as follows. Let $G(V, E)$ be a simple, undirected graph generated by the *random graph* model i.e. $G = G(n, p)$. Divide V into ℓ subsets of equal sizes i.e. $S = \{S_1, \ldots, S_\ell\}$ where each S_i, a *hidden partition*, is a disjoint set of vertices i.e. $S_i \cap S_j = \emptyset$ and $S_1 \cup \cdots \cup S_\ell = V$ and $|S_i| = m = n/\ell$. For each hidden partition, we add an *additional edge layer* by considering each as a random graph $G(m, q)$. If $q = 0$, then the graph G is just a normal random graph. For $q > 0$, the hidden partitions become denser and G has a more apparent cluster structure.

Assuming that $p = \Omega(1/n^{2/3})$ w.h.p. most vertices are in triangles, and if each partition is relatively dense then the edges with the most triangles should lie within the partitions rather than between them. Hence, if we apply k-MXT to the graph the vertices should be choosing k neighbours within their partition.

We show that, with $\ell = 4$, when $p \geq 3\sqrt{\log n/n}$ and $q = 2\beta p$ then for any $\beta > 0$ constant and $k \geq 1$, the probability that k-MXT wrongly chooses a between-partition edge tends to zero. However, this does not mean we can *automatically recover* the correct partitions, as the k-MXT fragments may be small. Experimental results show that with $k = 1$, the algorithm tends to break the partitions into many small fragments. Nevertheless, experimentally for $k \geq 2$, the algorithm is able recover the correct partitions provided β is a large enough constant.

Inspired by a study by Crandall, Backstrom, Huttenlocher, and Kleinberg (*Mapping the worlds photos*, [4]), and after some initial experiments, we noticed that, in some sense, a disk graph constructed using geographical co-ordinates of photographs show some structural resemblance to that of a planted ℓ partition in the sense that most photographs are taken at popular tourist spots. Even if the tourist venues are close, the edge density is still locally high. We experimented and evaluated the performance of the k-MXT algorithm in identifying popular venues. The algorithm used in [4] was mean-shift which under-performed at a detailed level, mainly because clusters are linear as people often take photos walking down the street. The DBSCAN algorithm [5] performed better, but has the advantage of two parameters (a disk size d and min-points m to include). By modifying k-MXT to only include those edges within a given disk d which induce at least m triangles, the algorithm can be adapted to outperform DBSCAN in some cases.

2   The k-MXT Algorithm

Let $G(V, E)$ be a simple, undirected graph. Let S be an empty graph, assign $V(S) = V(G)$ and $E(S) = \emptyset$. For each vertex $v \in V$, let $N(v)$ be its adjacency list, hence $d(v) = |N(v)|$ the degree of v. Each edge $(v, u) \in E(G)$ is given a weight which is the number of common neighbours of v and u, i.e. the cardinality of the set intersection of $N(v)$ and $N(u)$

$$w(v, u) = |N(v) \cap N(u)| = |\{w : w \in N(v) \text{ and } w \in N(u)\}|.$$

Next, for each vertex v we select a set of k incident edges of v which have the *highest weights*, let this set be $\triangle(v)$ (if there is a tie, edges are selected at random); if $d(v) < k$, then all incident edges are selected. Note that for each edge $(v, u) \in \triangle(v)$, the edge has a natural orientation, but we can choose to ignore its orientation. Finally, for each $v \in V$, the edge set $\triangle(v)$ is added to S.

0. Input: a simple graph, undirected graph $G = (V, E)$; number of edges to select (for each vertex) k.
1. Initialisation: an empty graph S, $V(S) = V(G)$ and $E(S) = \emptyset$.
2. For each edge $(i, j) \in E(G)$, compute $w(i, j)$.
3. For each vertex $i \in V(G)$, select $\triangle(i)$; for each edge $(i, j) \in \triangle(i)$: $E(S) = E(S) \cup (i, j)$.
4. Output: subgraph S.

3   The Planted ℓ Partitions Model

In the following sections we consider the case $\ell = 4$, thus $m = n/4$ is the size of each of the partitions P_1, P_2, P_3, P_4. We prove the following Theorem 1.

Theorem 1. *For $p \geq 3\sqrt{\log n/n}$ and $q = 2\beta p$ for any $\beta > 0$ constant and $k \geq 1$, then the probability that k-MXT wrongly chooses an inter-partition edge tends to zero as $O(1/n)$.*

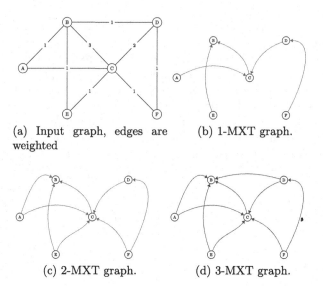

(a) Input graph, edges are weighted

(b) 1-MXT graph.

(c) 2-MXT graph.

(d) 3-MXT graph.

Fig. 1. An example of the k-MXT algorithm. Figure (a) presents the input graph with the weighted edges. Figures (b), (c) and (d) present the subgraph S whose values of k are $1, 2$ and 3, respectively.

By its nature k-MXT can only be used on graphs with triangles. The expected number of triangles in a random graph $G(n, p)$ is $E[\#\,\text{triangles}] = \binom{n}{3}p^3$. Let $p = n^{-c}$ this is $\Theta(n^{3-3c})$. If $c \leq 2/3$ then the expected number of triangles becomes at least linear with the graph size and hence significant globally. It can be shown that

Lemma 1. *Let* $G = G(n, p)$. *Let* X *be the expected number of triangles in* G *associated with vertex* v. *Let* $p = 1/n^{-c}$ *where* $c < 2/3$. *Then*

$$E[X] = \binom{n-1}{2}p^3 \approx \frac{n^{2-3c}}{2} \to \infty,$$

and

$$Pr(X = 0) = O(\frac{1}{n^{2-3c}} + \frac{1}{n^{1-c}}) \to 0.$$

From now on we assume $p = n^{-c}$ with $c < 2/3$.

3.1 The Expected Weight of an Intra/Inter Edges in the Planted ℓ Partitions Model

Let P be a partition. For $u, v \in P$, the edge (u, v) is an intra-partition edge. If $x \notin P$ then (u, x) is an inter-partition edge.

Consider a pair of vertices $v, u \in P_1$. The probability that the edge (v, u) is not present in the *global* graph G is $1 - p$. Further, the probability that the same edge is not present in the *hidden partition* P_1 is $1 - q$. Hence, let r be the probability that such edge exists, then provided that $pq = o(1)$

$$\Pr(\text{an intra edge } (v, u)) = r = 1 - (1 - p)(1 - q) = p + q - pq \approx p + q \quad (1)$$

Let $X_{vu} = X(intra, vu)$ be the number of vertices w with edges (w, u) and (w, v). For a pair of vertices v, x in *different partitions*, let $Y_{vx} = Y(inter, vx)$ be the number of vertices w with edges (w, v) and (w, x). We note that these variables do not depend on the presence or absence of the edge (v, u) or (v, x).

Lemma 2. *Let* $q = 2\beta p$. *Let* (v, u) *be an intra-partition edge and let* (v, x) *be an inter-partition edge. Then*

$$E[X(intra, vu)] = np^2(1 + \beta + \beta^2), \quad and \quad E[Y(inter, vx)] = np^2(1 + \beta).$$

Consequently, let $E[X(intra)] = \alpha E[Y(inter)]$ *then* $\alpha \approx 1 + \frac{\beta^2}{1+\beta}$.

The expected number of triangles on the intra-partition edges (v, u), or equivalently, the expected weight of an intra edge is, substituting $r \approx p+q = p(1+2\beta)$,

$$E[X(intra, vu)] = (m - 2)r^2 + 3mp^2 \approx m(r^2 + 3p^2) = np^2(1 + \beta + \beta^2). \quad (2)$$

For an inter-partition edge (v, x) in which $x \notin P_1$, e.g. $x \in P_2$, the probability that a triplet $\{v, x, y\}$ with y being in either P_1 or P_2 is rp. For any other vertex z that is not in the same partition as x and y, the triplet $\{v, x, z\}$ exists with probability p^2. Thus, the expected number of triangles on an inter edge (v, x) is

$$E[Y(inter, vx)] = 2(m - 1)rp + 2mp^2 \approx 2mp(r + p) = np^2(1 + \beta). \quad (3)$$

Putting $E[X(intra)] = \alpha E[Y(inter)]$, it follows that $\alpha \approx 1 + \beta^2/(1 + \beta)$. □

3.2 Threshold for β to Avoid Selecting Inter-cluster Edges

For convenience, denote the weight of an intra edge (v, u) and inter edge (v, x) by X and Y instead of $X(intra, vu)$ and $Y(inter, vx)$. Substituting $p = n^{-c}$ into Eq. (3) and (2) yields

$$E[Y] = n^{1-2c}(1 + \beta), \quad and \quad E[X] = n^{1-2c}(1 + \beta + \beta^2).$$

For $c > 1/2$, $E[Y] = O(n^{-c}) \to 0$ as $n \to \infty$. As this case is of little interest, we exclude it and assume $p > n^{-1/2}$.

For $c \leq 1/2$ we examine Y using the Hoeffding-Chernoff concentration inequality. Choose $\delta = \sqrt{\omega/E[Y]}$ with $\omega = 9 \log n$ (note that Hoeffding-Chernoff requires $0 < \delta < 1$, which will be examined below), we get

$$\Pr\left(|Y - E[Y]| \geq \delta E[Y]\right) \leq e^{-\delta^2 E[Y]/3} = e^{-\omega^2/3} = e^{-3 \log n} = n^{-3}. \quad (4)$$

As there are at most $\binom{n}{2}$ edges (v, x) to consider, the probability any edge (v, x) deviates is $O(n^2 \times n^{-3}) = O(1/n)$.

Similarly, we choose $\widehat{\delta} = \sqrt{\omega/E[X]}$. Then X is bounded as

$$\Pr\{|X - E[X]| \geq \widehat{\delta}E[X]\} \leq e^{-\widehat{\delta}^2 E[X]/3} = e^{-3\log n} = n^{-3}. \tag{5}$$

Hence the probability that any edge (v, u) (with v and u in the same partition) deviates is also $O(1/n)$.

It is seen that Y is unlikely to be more than $(1 + \delta)E[Y]$; and X is unlikely to be less than $(1 - \widehat{\delta})E[X]$. Therefore set

$$\delta E[Y] + \widehat{\delta}E[X] \leq E[X] - E[Y],$$

which simplifies to

$$\sqrt{\omega} = 3\sqrt{\log n} \leq \frac{E[X] - E[Y]}{\sqrt{E[X]} + \sqrt{E[Y]}} = .n^{1/2-c}\left(\sqrt{\beta^2 + \beta + 1} - \sqrt{\beta + 1}\right). \tag{6}$$

For $c < 1/2$. As $\widehat{\delta} < \delta$, Chernoff's inequality requires that $\delta = \sqrt{\omega/E[Y]} < 1$, hence

$$3\sqrt{\log n} < \sqrt{E[Y]} = n^{1/2-c}\sqrt{1 + \beta},$$

which implies that $p = n^{-c} \geq 3\sqrt{\log n/n}$. Thus (6) yields

$$3n^{c-1/2}\sqrt{\log n} < \sqrt{\beta^2 + \beta + 1} - \sqrt{\beta + 1}, \tag{7}$$

as $c < 1/2$ the left hand side is $O(n^{-\epsilon}) \to 0$ for $n \to \infty$. The right hand side is at least constant for $\beta > 0$. Thus (7) is true for any $\beta > 0$, and we have Theorem 1.

3.3   Experiments

We conducted experiments to investigate the accuracy of the algorithm when β increases. To generate the planted ℓ partitions graphs, we set: $n = 10^4$, $p = 1/\sqrt{n}$; $q \in [0, 14]$. For each unique set of parameters, we generated 10 different graphs, and fragment it with $k = 1, 2, 3, 4, 5$ which we abbreviate as $1, 2, 3, 4, 5$-MXT. The value of $p = 1/\sqrt{n}$ was chosen as it is the threshold for triangles to appear on the inter-partition edges, the expected number of triangles being constant (see (3)). To evaluate the accuracy of the partitions, we use the following metrics.

Adjusted Rand Index. The Adjusted Rand Index (ARI) [8] is a metric based on the principle of *pair counting*. It measures the similarity of two different sets by counting the number of pairs of vertices which are classified in the same and different clusters in the two given sets. In more detail, given a set, and two partitions derived from it: $X = X_1, ..., X_m$ and $Y = Y_1, ...Y_n$, where m and n may or may not be equal, the ARI measures the similarities of X and Y, yielding a score ranges from $[-1; 1]$. A higher score indicates high similarity with 1 being an exact match. In our case, X is the hidden partitions and Y is the components in the resulting k-MXT subgraph.

Fraction of Incorrect Edges. An *incorrect edge* is an edge selected by the k-MXT algorithm that connects a pair of vertices from different partitions. Let $P(v)$ be the index of the hidden partition which vertex v belongs. The edge (v, u), selected by the algorithm, is an incorrect edge if $P(v) \neq P(u)$. The *fraction of incorrect directed-edge* (FoID) is the ratio between the total number of *incorrect* edges over the total number of edges in subgraph H

$$S(v, u) = \begin{cases} 1 \text{ if } P(v) \neq P(u) \\ 0 \text{ otherwise} \end{cases}$$

then

$$\text{FoID} = \frac{\sum_{(v_i, v_j) \in E(H)} S(v_i, v_j)}{|E(H)|}$$

Thus, the value of FoID is in $[0; 1]$; a score of 1 indicates all edges are *incorrect*.

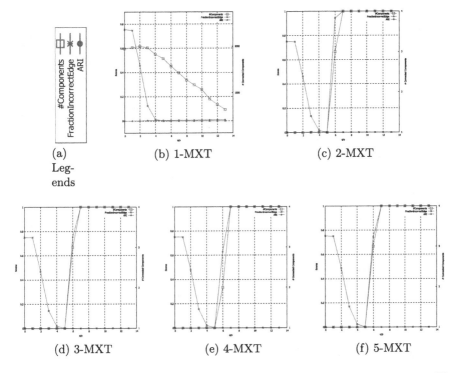

Fig. 2. The figures present the performance of k-MXT algorithm in experiments. The x-axis is the ratio $q/p = 2\beta$. The **primary y-axis** (left) corresponds to the scores for the *FoID (blue)* and *ARI (green)*. The **secondary y-axis** (right) corresponds to the *number of connected components (red)*. It can be seen that there is a difference between 1-MXT and the rest. (Color figure online)

Results. The results of the experiments are reported in Fig. 2. It can be seen that there is a difference between 1-MXT and the rest.

Particularly, for 1-*MXT*, although the FoID ≈ 0 i.e. the number of *incorrect edges* is approximately 0 (for $q/p \geq 6$) the accuracy measure ARI is also 0. This is because the 1-MXT *breaks the correct hidden partitions into many small fragments* (as indicated by the number of components). To achieve a high ARI scores, it is required that the number of components to be close to 4 i.e. the number of hidden partitions. Therefore, even though 1-MXT *does not select any incorrect* edge, its accuracy score is low. Picturesquely, 1-MXT produces *small fragments* which are contained inside the correct, hidden partitions.

It appears that experimentally $2, 3, 4, 5$-MXT *have a threshold at* $q/p \approx 8$, above which the algorithms find the correct partitions i.e. number of components is 4 and ARI scores are 1.

4   Application of k-MXT to Clustering Geo-Tagged Data

In this section, we consider the problem of clustering geospatial points, collected from *photographs* as a graph problem. Given a dataset consists of geographical coordinates i.e. latitude and longitude, we create a *proximity graph* as follows. For every point (x_v, y_v) where x_v and y_v are the latitude and longitude of vertex v, we connect v to every point u located within the disk radius d centered at v, i.e. $dist(v, u) \leq d$ where $dist(v, u)$ is a distance function.

The result is a simple, undirected graph, which captures the essential spatial information. Further, because the points are biased toward specific locations (popular landmarks, attractions, etc.), the graph reveals the structure of the underlying data (see Fig. 3). The subgraphs at dense regions of points (presumably popular attractions) are *densely* connected, with many triangles. Also, these dense regions are connected by sparser subgraphs whose edges subtend few triangles. We investigate the application of the k-MXT algorithm to identifying the denser regions.

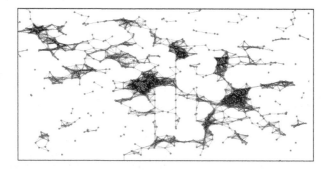

Fig. 3. An example of proximity graph consisting of $\approx$2,000 points. For each pair of vertices, we add an edge if the distance between them is at most $d = 100$ m.

4.1   Data Collection

We select Flickr as the main photo-sharing platform as it is easily accessible and possesses a large database of geo-tagged photos. To collect photographs, we identify a bounding rectangle covering a desired region. We choose the city of London. The bounding rectangle is further divided into (a large number of) sub-rectangles. Finally, for each sub-rectangle, we issue a query using Flickr's API to collect photographs which are taken inside it.

Using the data collected, we created two datasets. The *small set* consists of 4,000 data points within a bounding rectangle covering St. Paul's Cathedral and Tate Modern. The *large set* consists of 45,000 data points covering a larger region of central London.

4.2   Comparison of Algorithms

We compare the performance of k-MXT with the Mean-shift and DBSCAN algorithms, which we briefly outline below. Due to space restrictions, the complexity analysis of the algorithms and experimental runtime comparisons is given in the appendix.

Mean-Shift. Mean shift is a non-parametric technique for locating the maxima of a density function [2]. It is used as the method of choice for clustering in the paper [4]. Mean shift follows an iterative procedure to locate the mode of an underlying probability distribution given a set of sample points. In more detail, for each point p, let C be a circle with radius h with p being its centre. Using every point p_j located inside C, a *weighted mean* is calculated by

$$m(p_i) = \frac{\sum_{p_j \in C_i} \{p_j . f(p_j, p_i)\}}{\sum_{p_j \in C_i} f(p_j, p_i)}, \tag{8}$$

where $f(p_j, p_i)$ is a kernel function that determines the weight of nearby points for re-estimation of the mean; hence $p_j . f(p_j, p_i)$ multiplies the coordinate vector p_j with the scalar f. We use the Gaussian kernel. The initial data-point is then *shifted* to $m(p_i)$ i.e. $p_i \leftarrow m(p_i)$ and then continue recursively for p_i. The recursion stops if the mean converges, practically when $dist(m(p_i), p_i) \leq \lambda$; or until a maximum number of iteration T_{max} is reached. The points which share the same mode (or approximately close) is put in the same cluster.

DBSCAN. (Density-Based Spatial Clustering of Applications with Noise) [5] is a popular spatial clustering algorithm. The algorithm requires a distance parameter ϵ and an optional parameter *minPts*. Consider a data point p and a circle C with radius ϵ, centred at p. The point p is classified as a core point if and only if there are at least *minPts* number of neighbours points within C. A point q is *reachable* from p if there exists a path from p to q formed by $p_1, ..., p_n$ where $p_1 = p$ and $p_n = q$, in which p_i for $1 < i < n$ must be a core point in its own disc. The only exception is the final point q. All points *not reachable* are *noise*. Non-core points are points which are not core but *reachable* from another core point. Clusters are formed by grouping together *core* and *non-core* points such that in each cluster every point is reachable from the other. Note that non-core points cannot be used to *reach* others.

The Algorithm Parameters. To calculate distance, we treat the coordinates as points in two dimensional Euclidean space, because the errors are tolerable given the scale we are looking at. We consider three radiuses: *10* m, *25* m and *50* m. The other parameters are listed below

1. k-MXT: $k = 1, 2, 3, 4$;
2. DBSCAN: $minPts = 3, 20, 40, 80$;
3. Mean shift: $T_{\max} = 50$; $\lambda = 1$ (metre).

4.3   Results and Evaluation

Visualisation. For each cluster we draw a *convex hull* which is the minimum bounding polygon covering every point within that cluster. The basic parameter is the radius d, ϵ, h of the bounding circle in MXT, DBSCAN and Mean-shift respectively. To distinguish the clusters generated by different parameters k and $minPts$ of MXT and DBSCAN, we use different colours for the bounding polygons, see Table 1 for more detail. To demonstrate the change in cluster size obtained by altering the parameters k and $minPts$, we *overlayed* the resulting clusters.

The geographic location of the small dataset (between St Paul's and Tate Modern) is shown in this figure. The results of the small dataset are presented in Fig. 4. Results for the large dataset are presented in multiple figures in the Appendix.

Accuracy. We evaluate the accuracy of the algorithms by visual inspection. More specifically, we examine whether the polygons produced by the algorithms cover *meaningful* regions (see Fig. 11 for a map of London with some identified points of interest). Overall, DBSCAN and k-MXT appear to be the best algorithms that produce compact regions of points that cover some identifiable popular attractions in London. Moreover both algorithms produce polygons which follow the 'shape' of the density of points.

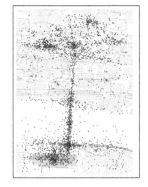

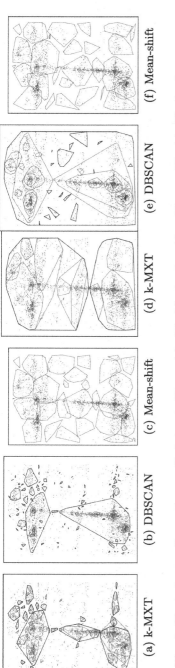

Fig. 4. Clustering results on the **small dataset**. Figure (a), (b), (c): $d, \epsilon, h = 10\,\mathrm{m}$. Figures (d), (e), (f): $d, \epsilon, h = 25\,\mathrm{m}$.

(a) k-MXT (b) DBSCAN (c) Mean-shift (d) k-MXT (e) DBSCAN (f) Mean-shift

Table 1. Colour coding schemes correspond to specific parameter and also the size of the resulting cluster.

Size	Corresponding parameter			Colour
	Mean shift	minPts (DBSCAN)	k (k-MXT)	
Large	50m	3pt	4	● Black
Medium	25m	20pt	3	● Red
Small	10m	40pt	2	● Blue
Very small		80pt	1	● Green

Table 2. Running time in *seconds*. Note: for k-MXT, *the pre-processing and graph construction time is excluded*.

		k-MXT	DBSCAN	MeanShift
Small set	10 m	0.03	0.02	180
	25 m	0.33	0.04	360
	50 m	0.47	0.05	600
Large set	10 m	0.56	0.38	
	25 m	8.55	1.35	
	50 m	61.5	4.12	

At a small radius ($d = 10\,\text{m}$), the polygons produced by $k = 1$ are too small and fragmented, those produced by $k = 2, 3, 4$ just about right on the small dataset (Figs. 4, 8), whereas $k = 2$ seems best on the large dataset (Figs. 9, 10, 11). At radius $d = 10\,\text{m}$ the results for k-MXT seem more pleasing, compared to DBSCAN.

At larger radius ($d = 25\,\text{m}$) DBSCAN improves on k-MXT (Figs. 12, 13). The polygons produced by $k \geq 3$ are rather large. However this can be overcome by truncating low weight edges (see Fig. 14, and the section weighted k-MXT below).

Thus 2-MXT at $d = 10\,\text{m}$, seems the best choice for geo-tagged clustering at city level. See in particular the London map produced by 2-MXT in Fig. 5.

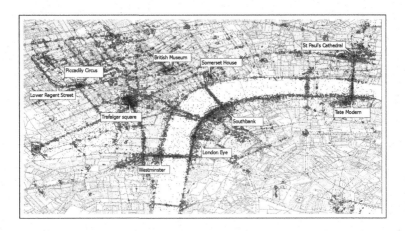

Fig. 5. Clusterings produced by 2-MXT ($d = 10\,\text{m}$), layered on London's map. Some popular tourist attractions are identified and labelled.

The polygons produced by DBSCAN and k-MXT differ in shape. Polygons produced by k-MXT are more rounded, whereas DBSCAN sometimes faithfully follows the street layout. Both algorithms produce distinct, well defined clusters; whereas, the polygons produced by *mean shift* seem similar and uninformative regardless of the distance h see Fig. 4(c) and (f).

A Measure of Density. The resulting polygons lead us to the question: What makes a cluster look better than the others? Consider the six figures show in Fig. 6. Which figure, visually, looks better compared to the other? The answer is, perhaps, (e) or (f). This is because the top polygons i.e. (a) (b) and (c) cover exceedingly large areas, thus there is a significantly amount of empty space in each figure. Whereas, polygon (d) (e) and (f) look much better, being able to closely fit the centre dense region. Although (f) is visibly the smallest, but it can be argued that it is too small as many of the relevant points lie outside the polygon. Hence, (e) is arguably better.

Motivated by this observation, we derive a measure of density as follows. Let $n(P)$ be the number of points within the polygon P, and $A(P)$ be its area, then the density $R(P)$ is defined by

$$R(P) = \frac{n(P)}{A(P)}. \tag{9}$$

Thus $R(P)$ measures the data density within a given region i.e. the number of contained data points per unit of area. In our experiments, $A(P)$ is typically measured in square metres m^2, thus $R(P)$ yields the number of points per m^2. A high score indicates a high level of density of *dots* within such polygon. However, it may not always indicate a better *cluster*.

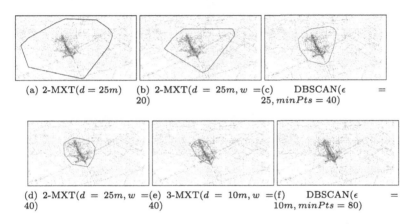

(a) 2-MXT($d = 25m$)   (b) 2-MXT($d = 25m, w =$ (c)   DBSCAN(ϵ =
                         20)   25, $minPts = 40$)

(d) 2-MXT($d = 25m, w =$ (e) 3-MXT($d = 10m, w =$ (f)   DBSCAN(ϵ =
40)   40)   10m, $minPts = 80$)

Fig. 6. Trafalgar Square. Which figure looks better? Note that the k-MXT algorithm used to produce figures (b), (d) and (e) has an additional parameter w which is introduced in Sect. 4.4.

Given the number of points and the areas of the region of our dataset, experimentally, we found that a density between $0.1 \leq R(P) \leq 0.5$ can be considered a good result. For instances, the density of the polygons in bottom row in Fig. 6(e) and (f) is 0.072, 0.101 and 0.291, respectively. Scores higher than 1 are often occur in very small polygons but having several points thus having an abnormally high score. We often exclude these cases.

4.4 Parameterising the k-MXT Algorithm

A property of the k-MXT algorithm is that when k increases the density of clusters produced by k-MXT is significantly reduced. This is because data points are not evenly distributed as they are biased toward specific locations (landmarks, attractions, etc.). Thus by increasing k, the area of the polygons are increased (as a vertex now select more edges), thus reducing the density of the polygons. This behaviour is might be due to the absence of a *noise controlling mechanism*.

The small polygons are considered as small clusters so far, not noise. Hence, to improve the k-MXT algorithm, we introduce a parameter w similar to the *minPts* parameter of DBSCAN.

The parameterised algorithm k-MXT(w_{min}) adds a simple *edge filter* to the graph construction process so that for each edge e if $w(e) < w_{\min}$ then delete e. Resulting vertices of degree zero are considered as *noise*.

Figure 7 plots selected clusters over the region of Trafalgar Square. These clusters are produced by k-MXT($w = 40, 80$) to compare directly with DBSCAN ($minPts = 40, 80$), with $d, \epsilon = 25$ m. Overall, it can be seen that the introduction of w significantly improves the results of k-MXT.

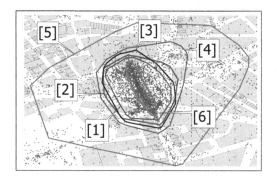

Fig. 7. Choice of region for geo-tags in Trafalgar Square.

The parameters and resulting density R for each polygon in Fig. 7 are:

- [1]-blue 2-MXT($d = 25$m, $w = 80$), $R = 0.128$;
- [2]-red 2-MXT($d = 25$m, $w = 40$), $R = 0.101$;
- [3]-black DBSCAN($\epsilon = 25$m, $minPts = 80$), $R = 0.101$;
- [4]-brown DBSCAN($\epsilon = 25$m, $minPts = 40$), $R = 0.072$;
- [5]-green 2-MXT($d = 25$m, $w = 0$), $R = 0.026$;
- [6]-turquoise DBSCAN($\epsilon = 10$m, $minPts = 80$), $R = 0.291$.

If we use 2-MXT($d = 25$m) on the large dataset, Fig. 12 shows the results, however visually it requires improvement. This is obtained by using parameterised 2-MXT. Figure 14 compares the weighted versions of DBSCAN and 2-MXT. Figures (a) and (d) seem best.

A   Appendix

A.1   Complexity Analysis

The k-MXT procedure can be broken down into two main tasks: *constructing the disk graph* and selecting the *k-Max Triangles neighbours* for each vertex.

Graph Construction. To construct the proximity graph, we need to find the points inside each vertex's disc. The *naive approach* takes $O(n^2)$ operations. A simple improvement is to separate the set of latitudes and longitudes and sort them. For each vertex $v = (x_v, y_v)$ and each coordinate x_v and y_v, we locate the values within a fixed distance from it using range binary search. The result, for *each coordinate*, is a set of points which are located within a fixed distance d from the queried point i.e. $S(x_v) = \{u : |x_u - x_v| \le d\}$ and $S(y_v) = \{w : |y_w - y_v| \le d\}$. The intersection of these two sets $S(x_v) \cap S(y_v)$ can be done using the smaller set, and yields the set of points bounded by a square of width $2d$ centered at v. To transform the *bounding square* to a *bounding circle* then requires an additional computation step. Overall, the complexity of the *improved naive method* takes

$$\underbrace{O(n \log n)}_{\text{sort}} + \underbrace{O(n \log n)}_{\text{range search}} + \underbrace{O(n \times \min_{v \in V}\{|S(x_v)|, |S(y_v)|\})}_{\text{connect edges}}.$$

Further improvement requires using spatial data structures such as R-trees or kd-trees. The construction of such trees take, on average, $O(n \log n)$. A search query takes $O(\log n)$ on average and $O(n)$ worst. Also, note that both structures *search operation* only support query by *rectangle*. Thus, an additional step is required to locate points within a vertex's *circle* of radius d.

We give an experimental running time of the graph construction methods in Table 3. We used the C++ Boost Geometry library [11] for an implementation of R-tree and the Approximate Nearest Neighbour (ANN) for kd-tree.

Additionally, there is further consideration of calculating distances using great-circle distance when carrying out geo-tagged clustering at country or world scale. Note that only the R-tree implementation supports this operation. Our experiments show that the *improved naive* method out-performs R-trees (Table 3).

Selecting Neighbours. Given the constructed graph, to select the k-Max Triangle edges for each vertex requires,

1. For each edge: calculate the number of common neighbours;
2. For each vertex: select the k highest scores;
3. Find the connected components of the resulting graph fragments.

If the adjacency lists are sorted (implying $O(n \times d_{max} \log d_{max})$ *preprocessing*), the *first task* is equivalent to finding the set intersection. Thus for each edge it takes at most $d(u) + d(v)$. Hence, the overall computation cost is

$$\sum_{v \in V} \sum_{u \in N(v)} (d(v) + d(u)) = \sum_{v \in V} d^2(v) + \sum_{u \in N(v)} \sum_{v \in V} d(u) \le 2d_{max} \sum_{v \in V} d(v) = 4|E|d_{max}.$$

Thus the first task takes $O(|E| \times d_{max})$.

Table 3. Large dataset: $n = 45,000$. Graph construction time in *seconds* and averaged over 10 executions. The best running time are highlighted. Interestingly, the *improved naive* method performs better than the R-tree when computing the *spherical* coordinates.

	Naive		Improved Naive		R-tree		kd-tree
	Cartesian	Spherical	Cartesian	Spherical	Cartesian	Spherical	Cartesian
10 m	82.1	419.4	1.6	**5.6**	1.3	9.1	**1**
25 m	82.1	419.4	3.6	**13.59**	3.1	26.7	**1.5**
50 m	82.1	419.1	6.9	**26.35**	7.3	65.3	**2.6**

Table 4. Table presents the average density and the density of the top polygons for the large dataset with $d, \epsilon = 25$ m. Results for the 2-MXT algorithm ($w = 40, 80$) and DBSCAN ($minPts = 40, 80$) are also included.

	Density	2-MXT ($w = 40$)	2-MXT ($minPts = 40$)	DBSCAN ($minPts = 40$)	2-MXT ($w = 80$)	DBSCAN ($minPts = 80$)
Small Area: $[100, 1000]\, m^2$	Best	0.163	0.215	0.052	0.274	NA
	Average	0.055	0.08	0.045	0.12	NA
Medium Area: $[1000, 5000]\, m^2$	Best	0.05	0.103	0.044	0.113	0.068
	Average	0.013	0.052	0.031	0.086	0.054
Large Area: $> 5000\, m^2$	Best	0.04	0.101	0.072	0.128	0.101
	Average	0.01	0.068	0.034	0.128	0.062

The *second task* is done using a priority queue i.e. min-heap which takes at most $\sum_{v \in V} d(v) \log k = \log k \times 2|E| = O(|E|)$, for fixed k.

The *final task* is to compute the connected components of the k-MXT subgraph. This can be done using any classical algorithm in linear time in the number of edges in the component, $O(kn)$ overall. For small values of k this is $O(n)$.

Overall, the *fragmenting process* has a running time of

$$\underbrace{O(n \times d_{\max} \log d_{\max})}_{\text{pre-processing}} + \underbrace{O(|E|d_{\max})}_{\text{set intersect}} + \underbrace{O(|E|)}_{k\text{-max scores}} + \underbrace{O(n)}_{\text{components}} = O(|E|d_{\max}).$$

In comparison, DBSCAN implemented with R-trees or kd-trees has an average complexity of $O(n \log n)$ [5]. For mean shift, a loose theoretical running time is $O(n \times T_{\max})$ where $T_{\max}$ is the maximum number of iterations allowed for each query. In practice, we usually set the distance to determines convergence to $\lambda = 0.5\,\mathrm{m}$, and noticed that the mode converged in relatively fewer iterations than T_{max}.

Experimental Running Time. DBSCAN is executed using the R package dbscan [10], a *fast re-implementation* of the original algorithm in C++. Mean-shift is executed using the R package MeanShift [9].

Table 2 presents the experimental running time of the algorithms. The results of the small dataset experiments show the mean shift has the slowest running time; hence it was excluded in the large experiment. Furthermore, in both experiments, optimized DBSCAN outperformed *the current implementation* of k-MXT, which is hardly surprising. For k-MXT, it is seen that the running time for the *clustering procedure* seems to scale quadratically with the distance, which determines the number of edges in the graph hence d_{max}. This is probably due to the $O(|E|d_{max}) \approx O(n(d_{max})^2)$ set intersection.

k-MXT applied to the small dataset *d* = 10m.

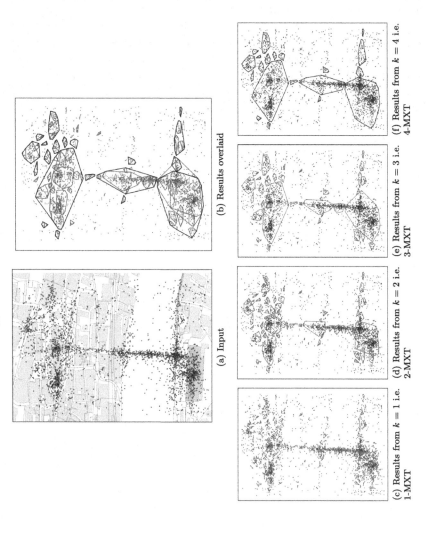

(a) Input

(b) Results overlaid

(c) Results from *k* = 1 i.e. 1-MXT

(d) Results from *k* = 2 i.e. 2-MXT

(e) Results from *k* = 3 i.e. 3-MXT

(f) Results from *k* = 4 i.e. 4-MXT

Fig. 8. *k*-MXT algorithms on the *small dataset* with *d* = 10 m and *k* = 1, 2, 3 and 4. For *k* ≥ 2, some regions are clearly defined. One can roughly identify some straight lines which are bridges, roads or footpaths. Notice that the *polygons* expand with *k*.

k-MXT on large dataset, $d = 10$m.

Fig. 9. k-MXT($d = 10$)m. The results for each k have been overlaid on the same layer, demonstrating the expanding regions as k is increased. The region in the top right hand corner is the **small dataset**, shown in previous Figs. 8 and 4.

DBSCAN on large dataset, ϵ = 10m.

Fig. 10. DBSCAN(ϵ = 10 m). As $minPts$ is increased, the number of clusters is decreased as the area of the polygons also decrease. There is no visible green (correspond to $minPts$ = 80) polygon, few of the blue polygons ($minPts$ = 40), while the area of the red polygon ($minPts$ = 20) are relatively small. (Color figure online)

Clusterings produced by 2-MXT($d = 10m$), shown on map.

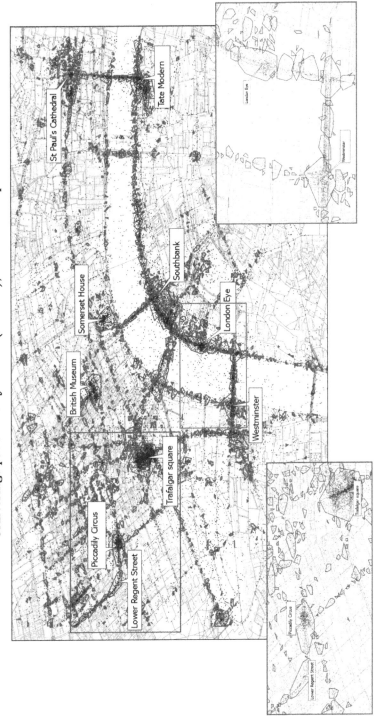

Fig. 11. 2-MXT($d = 10$ m), layered on London's map. Some popular tourist attractions are identified and labelled. It can be seen that there are clusters/polygons corresponding to these places. The two separated pictures provide a closer look at the two regions, specified by its border's colour. (Color figure online)

k-MXT on large dataset, $d = 25$m.

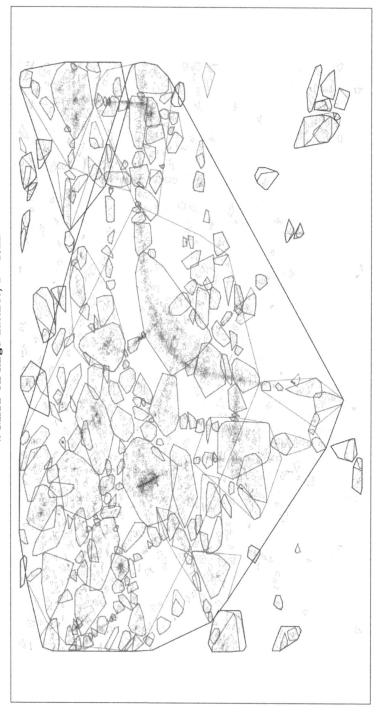

Fig. 12. k-MXT($d = 25$ m). The green (smallest, $k = 1$) polygons are very small and its orientation seems quite arbitrary. The blue ($k = 2$) polygons have expanded considerably but still retain some interests. While the red ($k = 3$) and black ($k = 4$) regions cover very large areas of central London. (Color figure online)

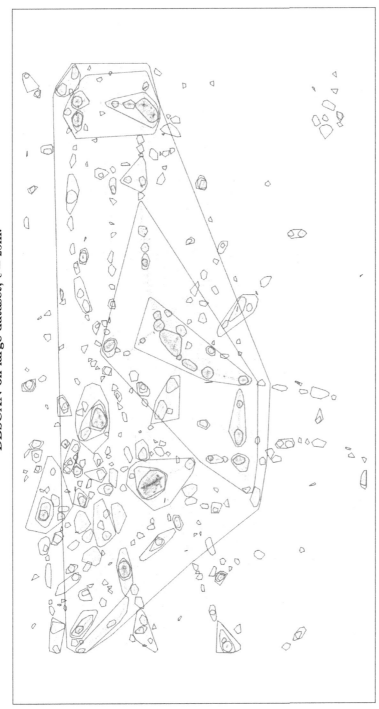

DBSCAN on large dataset, $\epsilon = 25$m.

Fig. 13. DBSCAN($\epsilon = 25$ m). Overall, the picture shows a decent level of clarity. Compare to *k*-MXT in Fig. 12, the areas of regions produced are much smaller. Notably, the green (small, $minPts = 80$) and red (medium, $minPts = 40$) polygons cover some interesting regions. (Color figure online)

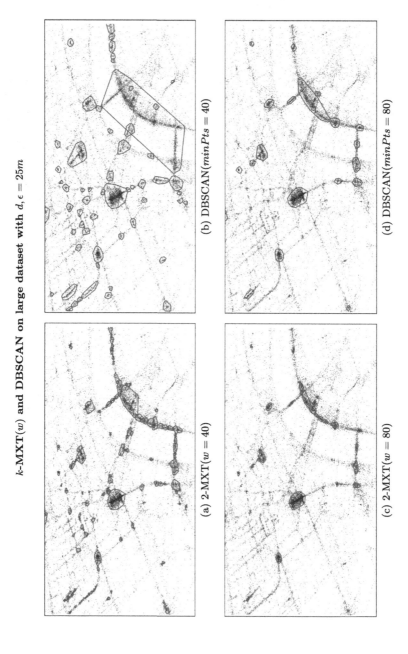

Fig. 14. Parameterized *k*-MXT and DBSCAN comparison. Truncating low weight edges in *k*-MXT improves performance

References

1. Bollobas, B.: Random Graphs. Cambridge Studies in Advanced Mathematics, 2nd edn. Cambridge University Press, Cambridge (2001)
2. Cheng, Y.: Mean shift, mode seeking, and clustering. IEEE Trans. Pattern Anal. Mach. Intell. **17**(8), 790–799 (1995)
3. Condon, A., Karp, R.M.: Algorithms for graph partitioning on the planted partition model. Random Struct. Algorithms **18**(2), 116–140 (2001)
4. Crandall, D.J., Backstrom, L., Huttenlocher, D., Kleinberg, J.: Mapping the world's photos. In: Proceedings of the 18th International Conference on World Wide Web, WWW 2009, pp. 761–770. ACM, New York (2009)
5. Ester, M., Kriegel, H.P., Sander, J., Xu, X.: A density-based algorithm for discovering clusters in large spatial databases with noise, pp. 226–231. AAAI Press (1996)
6. Frieze, A., Michal, K.: Introduction to Random Graphs. Cambridge University Press, Cambridge (2015)
7. Girvan, M., Newman, M.E.J.: Community structure in social and biological networks. Proc. Natl. Acad. Sci. USA **99**, 7821–7826 (2002). 2001
8. Hubert, L., Arabie, P.: Comparing partitions. J. Classif. **2**(1), 193–218 (1985)
9. Ciollaro, M., Wang, D.: Package: MeanShift. https://cran.r-project.org/web/packages/MeanShift/MeanShift.pdf. Accessed 2017
10. Hahsler, M., et al.: Package: dbscan. https://cran.r-project.org/web/packages/dbscan/dbscan.pdf. Accessed 2017
11. Various. Boost.Geometry. http://www.boost.org/doc/libs/1.61.0/libs/geometry/doc/html/index.html. Accessed 2017

A Statistical Performance Analysis of Graph Clustering Algorithms

Pierre Miasnikof[1(✉)], Alexander Y. Shestopaloff[2], Anthony J. Bonner[3], and Yuri Lawryshyn[1]

[1] Department of Chemical Engineering and Applied Chemistry,
University of Toronto, Toronto, Canada
`p.miasnikof@mail.utoronto.ca`
[2] The Alan Turing Institute, London, UK
[3] Department of Computer Science, University of Toronto, Toronto, Canada

Abstract. Measuring graph clustering quality remains an open problem. Here, we introduce three statistical measures to address the problem. We empirically explore their behavior under a number of stress test scenarios and compare it to the commonly used modularity and conductance. Our measures are robust, immune to resolution limit, easy to intuitively interpret and also have a formal statistical interpretation. Our empirical stress test results confirm that our measures compare favorably to the established ones. In particular, they are shown to be more responsive to graph structure, less sensitive to sample size and breakdowns during numerical implementation and less sensitive to uncertainty in connectivity. These features are especially important in the context of larger data sets or when the data may contain errors in the connectivity patterns.

1 Introduction

While there are many graph clustering[1] algorithms in the literature (e.g., [15,17, 21,24]), measuring their performance, that is assessing the quality of the clusters they identify, remains an open problem [1,3,6,11–13,16,23]. Graph clustering is a form of unsupervised learning, where one typically cannot count on labeled data to assess results. For example, in [20], the authors correctly assert that "(...) *running a clustering algorithm over a set of randomly generated data points will always produce clusters which, however, have little meaning.*" Therefore, our only quality measure is a thorough assessment of the graph's and resulting clusters' connectivity patterns.

In this article, we present new clustering performance measures to assess the strength of the clustering returned by a specific algorithm and compare clusterings across algorithms on a specific graph. We restrict our attention to undirected

---

[1] Note on vocabulary: Although there are subtle differences between the concepts of graph clustering and community detection, in this document we use the two interchangeably.

© Springer International Publishing AG, part of Springer Nature 2018
A. Bonato et al. (Eds.): WAW 2018, LNCS 10836, pp. 170–184, 2018.
https://doi.org/10.1007/978-3-319-92871-5_11

unweighted and weighted graphs, with no self-loops or multiple edges. We begin with a review of two of the most common clustering performance measures, modularity and conductance. We empirically demonstrate how these measures may, in some cases, be drowned out by graph structure and lack sensitivity. We also offer three alternative measures, which are shown to be more robust.

2 Performance Measures

In this section, we describe the two most popular performance metrics in the literature, namely modularity and conductance. We also present our own statistical measures, the "Kappas". In the following sections, we will empirically analyze their strengths and weaknesses.

2.1 Modularity

Modularity (Q) is by far the most popular measure of clustering performance [4,5,8,13,17–19]. It was originally introduced by Newman and Girvan in 2004 [17] and has been extensively used both as a performance measure and objective function for clustering algorithms (e.g., [2,7,17]). In this section, we present modularity (Q) as defined in [5].

$$Q = \sum_{i=1}^{k} \left(\underbrace{e_{i,i} - a_i^2}_{q_i} \right)$$

Where,

$$e_{i,i} = \frac{1}{2m} \sum_{v,w} A_{v,w}\, \delta(c_v, i)\delta(c_w, i)$$

$$a_i = \frac{1}{2m} \sum_{v} A_{v,.}\, \delta(c_v, i)$$

Here, $m = |E|$ is the total number of edges in the graph, k is the number of clusters, $A_{v,w}$ is the element at the intersection of the v-th row and w-th column of the adjacency matrix, $A_{v,.}$ is the entire v-th row of the adjacency matrix, $\delta(x, y)$ is the Kroenecker delta function, $e_{i,i}$ is the portion of vertex degree connecting vertices within cluster i, a_i is the total vertex degree in cluster i and c_v is the cluster in which vertex 'v' is clustered into by the algorithm. Putting it together, we get

$$Q = \sum_{i=1}^{k} \left[\underbrace{\frac{1}{2m} \sum_{v,w} A_{v,w}\, \delta(c_v, i)\delta(c_w, i)}_{e_{i,i}} - \underbrace{\frac{1}{4m^2} \left(\sum_{v} A_{v,.}\, \delta(c_v, i) \right)^2}_{a_i^2} \right]. \quad (1)$$

(A high value indicates densely connected clusters.)

In closing, it should be noted that modularity suffers from a resolution limit, as described by Fortunato and Bathélemy [9]. These authors describe how any (clustering) quality function that is defined as a sum of qualities of individual clusters where terms from smaller clusters are dominated by terms from larger clusters suffers from resolution limit. Because the smaller clusters' contribution to the sum is dominated by the larger clusters, the final result is also dominated and does not always reflect the structure accurately. Indeed, in (1), we see how larger clusters dominate the outer summation.

2.2   Conductance

Conductance (ϕ, Φ) is another popular clustering performance measure [6,13,14, 22,23]. In this article, we use the definition presented in [22].

At the individual cluster level,

$$\phi(S) = \frac{\partial(S)}{\min\left(d(S), d(V \setminus S)\right)}$$

At the graph level,

$$\Phi(G) = \min_{S} \phi(S)$$

Here, $\partial(S)$ is the number of edges joining vertices in cluster S to vertices outside S, $d(S)$ is the sum of vertex degrees within S and $d(V \setminus S)$ the sum of vertex degrees on the graph, outside S. (A low conductance indicates strongly connected clusters.)

2.3   The Kappas

Our overarching goal in developing our measures is to gauge the strength of connectivity on the graph in general, within individual clusters and between clusters. While the established measures of clustering strength, modularity and conductance, measure intra-cluster connectivity strength, we seek to measure the strength of intra- and inter-cluster connectivity relative to the overall graph's connectivity. For example, in a densely connected graph we expect clusters to be even more strongly connected and strong inter-cluster connections can be consistent with a good partition. Conversely, in a densely connected graph, poorly connected clusters or strong inter-cluster connectivity are symptoms of a poor clustering.

We define $\bar{K}$ as the graph's overall connectivity ratio, $\bar{K}_{\mathrm{intra}}$ as the measure of intra-cluster connectivity and $\bar{K}_{\mathrm{inter}}$ as the measure of inter-cluster connectivity. According to every definition of a good clustering, we expect that an efficient clustering algorithm will label vertices such that intra-cluster connectivity is greater than inter-cluster connectivity [8,18,19] (if the graph does indeed have a clustered structure). Under our model, we expect that a good clustering will group vertices so they form clusters whose vertices are more densely connected than the average connection between any two vertices on the graph. Similarly, we expect that a good clustering will group vertices so they form clusters whose vertices are less densely connected to those in other clusters than the average connection between any two vertices on the graph. In summary, we expect that under a good clustering the inequalities $\bar{K}_{\text{intra}} > \bar{K} > \bar{K}_{\text{inter}}$ will hold. Our model also allows these inequalities to be formulated as a hypothesis test, as will be shown later.

Below, we present the formulation for our clustering measures, for an unweighted undirected graph, but our metrics easily generalize to weighted graphs as well. In our formulation, we use the following variables: The set of all clusters is $C = \{C_1, \ldots, C_k\}$, with $|C| = k$, the total number of vertices in the graph is N, the total number of vertices in cluster i is $|C_i| = n_i$, the set of all edges on the graph is $E = \{e_1, \ldots, e_m\}$, where $|E| = m$. Finally, $E_{i,j}$ is the set of edges connecting a vertex in cluster i to a vertex in cluster j, and $|E_{i,j}| = m_{i,j}$. As a special case, note that $E_{i,i}$ is the set of edges within cluster i, and $m_{i,i}$ is the number of edges connecting vertices within cluster i.

In order to gauge the strength of the entire graph's, of each cluster's and each inter-cluster pair's connectivity, we take the ratio of the observed edges over the maximum possible number of edges given the number of vertices. For inter and intra cluster connectivity, we compute the ratio for each cluster or pair of clusters and take their mean as a graph-wide measure. All our measures lie in the $[0, 1]$ interval, with high values denoting highly connected graphs, clusters or cluster pairs and vice-versa.

We define the graph's connections ratio as

$$\bar{K} = \frac{|E|}{0.5 \times N(N-1)}.$$

The graph's connection ratio is the ratio of the total number of edges over the number of edges in a complete graph with the same number of vertices. The closer $\bar{K}$ is to 1, the closer the graph is to being a complete graph. Conversely, the closer $\bar{K}$ is to 0, the closer the graph is to being a set of disconnected vertices.

We also define the mean intra-cluster connections ratio as

$$\bar{K}_{\text{intra}} = \frac{1}{K} \sum_{i=1}^{\kappa} \frac{|E_{i,i}|}{0.5 \times n_i(n_i-1)}.$$

The mean intra-cluster connections ratio is the mean ratio of the number of edges within each cluster over the maximum number of edges that could possibly connect the vertices of each cluster. Each term in the summation is a measure of how closely each cluster is to being a clique. Each always lies on the interval $[0, 1]$, with a value of 0 indicating a cluster is just a set of disconnected vertices and a value of 1 indicating that a cluster is a clique. At the aggregate level, $\bar{K}_{\text{intra}}$ is the sample mean of the individual terms and also lies in the interval $[0, 1]$. Values close to 0 indicate poorly connected clusters on average, while values closer to 1 indicate densely connected clusters on average.

Finally, we define the mean inter-cluster connections ratio as

$$\bar{K}_{\text{inter}} = \frac{1}{0.5 \times k(k-1)} \sum_{i=1}^{k} \sum_{j=i+1}^{k} \frac{|E_{i,j}|}{0.5 \times ((n_i + n_j)(n_i + n_j - 1) - n_i(n_i - 1) - n_j(n_j - 1))}.$$

The mean inter-cluster connections ratio is the mean ratio of the number of edges joining vertices in two different clusters, over the total number of edges that could possibly connect each pair of vertices in each cluster pair (c_i, c_j). Each term in the double summation is a measure of how closely two clusters 'i' and 'j' are from forming a single clique. Each of these terms also lies in the interval $[0, 1]$, with a value of 0 indicating no connection between a pair of clusters and a value of 1 indicating that the pair of clusters forms a clique. At the aggregate level, $\bar{K}_{\text{inter}}$ is the sample mean of the individual terms of the summation and also lies in the interval $[0, 1]$. Values close to 0 indicate poor inter-cluster connections, on average, a desirable feature indicating strong cluster partitions, On the other hand, values closer to 1 indicate improperly partitioned clusters, on average.

It should also be mentioned that in cases where the connectivity patterns of the clusters is very noisy, the median of the summation terms can be used in lieu of the mean, in order to produce more robust measures. Unfortunately, this substitution makes statistical interpretation and significance testing less obvious.

Resolution Limit and Sensitivity to Cluster Size. It is important to note that neither $\bar{K}_{\text{intra}}$ nor $\bar{K}_{\text{inter}}$ are affected by individual cluster size and do not suffer from the resolution limit observed in modularity [9]. Large clusters do not skew their values, since all terms in the sums are scaled by the total number of possible edges within each cluster or pair of clusters and always lie on the $[0, 1]$ interval. This feature makes these measures robust to large "mega-clusters" that are often observed in real-world networks and to the fallacious tendency of clustering algorithms to lump all vertices together in a few very large clusters. (Naturally, $\bar{K}$ is a graph-wide measure that remains completely agnostic to clusters and their respective sizes.)

The equal weight carried by each cluster or pair of cluster does, however, have its drawbacks. Because our measures are unweighted means, they are somewhat sensitive to outliers. For example, a few unrepresentative small clusters could indeed skew the measures. However, the effect of outliers is typically smoothed out by the mean or can be corrected by the use of a weighted mean.

Statistical Interpretation of the Kappas. The main strength of our Kappas comes from their statistical definition. In the unweighted case, $\bar{K}$ is the probability any two nodes are connected, and in the weighted case it becomes the mean edge weight. Similarly, $\bar{K}_{\text{intra}}$ ($\bar{K}_{\text{inter}}$) is the mean probability two nodes within a cluster (between clusters) are connected or the mean intra-cluster (inter-cluster) edge weight.

In probabilistic terms, we expect a good clustering to partition the graph such that the probability there exists an edge $(e_{i,j})$ between two arbitrary nodes

'i' and 'j' is lower than the probability a connection exists if these nodes are in the same cluster (i.e., if $c_i = c_j$) and higher than when they belong to different clusters (i.e., $c_i \neq c_j$). Mathematically, we expect the following to hold:

$$Pr\,[e_{i,j}|c_i = c_j] > Pr[e_{i,j}] > Pr[e_{i,j}|c_i \neq c_j]$$

In the case of a weighted graph, these probabilities become expected values of edge weights between arbitrary vertices, vertices within and vertices between clusters, and we expect the following inequalities to hold:

$$E[e_{i,j}|c_i = c_j] > E[e_{i,j}] > E[e_{i,j}|c_i \neq c_j]$$

Defining our measures in this way, as estimates of an unknown "true" parameter, with an associated standard error, allows formal significance testing using a simple t-test. Such tests can be used to determine if the clusters identified by an algorithm are statistically significant. If they are, we expect the inequalities $\bar{K}_{intra} > \bar{K} > \bar{K}_{inter}$ to hold at a reasonable significance level. These inequalities are necessary and sufficient to conclude the clusterings returned by an algorithm are statistically (on average) consistent with the universally accepted definition of a clustering. [8,18,19]. Our statistics can also be used when comparing two or more algorithms' performances on a given graph. In such a case, in order to conclude algorithm 'a' is better than algorithm 'b', we should observe $\bar{K}^a_{intra} > \bar{K}^b_{intra}$ and $\bar{K}^a_{inter} < \bar{K}^b_{inter}$.

Finally, let us note that our statistical (i.e., non deterministic) definition also allows for uncertainty in the connectivity, another open problem [10]. Unlike modularity and conductance, our measures are defined as statistical measurements with associated standard errors, not deterministic quantities.

To formally confirm statistical significance and the strength with which the sufficient conditions are met, we formulate an appropriate null hypothesis and apply the t-test. Examples of such a test are shown in Sect. 4.4.

3   Computational Experiments

In order to empirically assess the accuracy of the various performance measures, to study their response to various graph structures and cluster labelings, we subject them to a number of numerical stress test scenarios, using simulated graphs and labels. The full experimental set-up of our individual tests and scenario details are described in the next sub-section.

Overall, our goal is to test the accuracy and robustness of our clustering measures and compare their behavior to that of the two main clustering measures in the literature (modularity and conductance). Simulation is used to generate test scenarios where the clustering structure is known in advance and could be modified easily. These test scenarios are then used to examine and compare the sensitivities of the kappas, modularity and conductance. Our scenarios include a number of contrived instances, which are useful to stress test our metrics through "extreme" examples and compare their behavior to those of the more established measures.

The overarching logic guiding our tests is that a good measure of inter- or intra- cluster connectivity should accurately reflect the simulated graph's structures. We would expect measures of intra-cluster connectivity, K_{intra} and modularity to increase in step with the simulated graph's connectivity levels, while we would expect conductance to display the inverse behavior. We would also expect K_{inter} to follow the fluctuations of inter-cluster connectivity.

It should also be mentioned that some authors have used so-called "ground-truth" data sets, networks where the nodes' cluster memberships were labeled, as benchmarks for clustering algorithm performance (e.g., [16,23,24]). Our approach is more general, data set and objective function independent. Arguably, the fact that an algorithm anecdotally provided accurate clustering on a labeled instance is no guarantee it will perform equally well on another (likely unlabeled) instance. In addition, our experiments provide us with an understanding of each measure's sensitivity and response to graph structure.

3.1   Experimental Set-Up

In the first set of experiments, shown in Table 1, we examine the effect of intra-cluster connectivity. We begin with a graph with no edges between any of the vertices and gradually increase intra-cluster connectivity in steps of 25%, while maintaining inter-cluster connectivity at 0% (e.g., 25% of nodes are connected to another node within their assigned cluster, 75% of nodes in each cluster have no connections at all, nodes with connections only have connections to other nodes within their assigned cluster, each cluster is disconnected from the rest of the graph).

We then examine the effect of inter-cluster connectivity on each measure. We begin with no inter-cluster connectivity and then increase it in steps of 25% (e.g., 25% of nodes are connected to 25% of nodes outside their cluster), while keeping intra-cluster connectivity at 0%. In other words, clusters are just sets of disconnected vertices. In these scenarios, we imagine an algorithm, one with a very poor cluster detection ability, that groups disconnected vertices into clusters with different levels of inter-connection to other clusters but with an intra-cluster connectivity that remains constant at 0%. Results are shown in Table 2.

In our experiments, we expect $\bar{K}_{intra}$ to increase in step with intra-cluster connection percentage. We also expect $\bar{K}_{inter}$ to increase in step with inter-cluster connection percentage. If this in-step increase occurs, it indicates our measures accurately detect the graph's connectivity structure.

Finally, in order to assess our measures' robustness, we repeat all the tests described above, but with the introduction of "noise" in the connectivity patterns. Noise is introduced in the form of 100% intra-(inter-) cluster connectivity. Results are shown in Tables 3 and 4.

In the tables that follow, we also report each graph's characteristics, for each experiment. The total number of vertices is denoted by N, the total number of clusters by $|C|$, and the total number of edges by $|E|$.

Table 1. Varying intra-cluster connectivity, no noise from inter-cluster connectivity

Pct Inter = 0, Pct Intra varies					
Pct Intra	0	25	50	75	100
N	10,048	9,440	9,666	10,493	10,039
$|C|$	200	200	200	200	200
$|E|$	0	76,942	160,147	269,341	336,942
$\bar{K}$	0.00	0.00	0.00	0.00	0.01
$\bar{K}_{intra}$	0.00	0.26	0.50	0.75	0.99
Std Err $(\bar{K}_{intra})$	0.00	0.01	0.01	0.01	0.01
$\bar{K}_{inter}$	0.00	0.00	0.00	0.00	0.00
Std Err $(\bar{K}_{inter})$	0.00	0.00	0.00	0.00	0.00
Φ	0.00	0.00	0.00	0.00	0.00
Q	0.00	0.99	0.99	0.99	0.99

Table 2. Varying inter-cluster connectivity, no noise from intra-cluster connectivity

Pct Intra = 0, Pct Inter varies					
Pct Inter	0	25	50	75	100
N	10,530	10,089	9,354	10,028	10,829
$|C|$	200	200	200	200	200
$|E|$	0	3,058,924	10,753,463	27,815,367	58,250,108
$\bar{K}$	0.00	0.06	0.25	0.55	0.99
$\bar{K}_{intra}$	0.00	0.00	0.00	0.00	0.00
Std Err $(\bar{K}_{intra})$	0.00	0.00	0.00	0.00	0.00
$\bar{K}_{inter}$	0.00	0.06	0.24	0.55	1.00
Std Err $(\bar{K}_{inter})$	0.00	0.00	0.00	0.00	0.00
Φ	0.00	1.00	1.00	1.00	1.00
Q	0.00	−0.01	−0.01	−0.01	−0.01

4   Discussion

As shown in Sect. 3, our "Kappas" behave as expected, even when subjected to noise. In all instances where the labeling of clusters reflects a good partition, the inequalities $\bar{K}_{\text{intra}} > \bar{K} > \bar{K}_{\text{inter}}$ hold and they do not hold in instances where the partition reflects poor clustering. For example, in Table 3, all instances are cases of poor clustering, by design. Similarly, in Table 4, instances where the percentage of inter-cluster connectivity is below 75% are examples designed to show good clustering and our inequalities hold in each.

More importantly, our inter- and intra-cluster measures follow the fluctuations of the graph's connectivity patterns more accurately than either modularity

Table 3. Varying intra-cluster connectivity, with noise from inter-cluster connectivity

Pct Inter = 100, Pct Intra varies					
Pct Intra	0	25	50	75	100
N	10,048	10,096	10,526	10,115	10,182
$|C|$	200	200	200	200	200
$|E|$	50,142,540	50,712,690	55,215,342	51,067,113	51,831,471
$\bar{K}$	0.99	1.00	1.00	1.00	1.00
$\bar{K}_{intra}$	0.00	0.25	0.50	0.74	0.98
Std Err (intra)	0.00	0.00	0.01	0.01	0.01
$\bar{K}_{inter}$	1.00	1.00	1.00	1.00	1.00
Std Err (inter)	0.00	0.00	0.00	0.00	0.00
Φ	1.00	1.00	1.00	0.99	0.99
Q	−0.01	0.00	0.00	0.00	0.00

Table 4. Varying inter-cluster connectivity, with noise from intra-cluster connectivity

Pct Intra = 100, Pct Inter varies					
Pct Inter	0	25	50	75	100
N	9,917	9,662	10,512	10,043	10,151
$|C|$	200	200	200	200	200
$|E|$	314,102	3,127,922	13,942,175	28,187,302	51,516,325
$\bar{K}$	0.01	0.07	0.25	0.56	1.00
$\bar{K}_{intra}$	1.00	0.99	1.00	1.00	1.00
Std Err (intra)	0.00	0.01	0.00	0.00	0.00
$\bar{K}_{inter}$	0.00	0.06	0.24	0.54	1.00
Std Err (inter)	0.00	0.00	0.00	0.00	0.00
Φ	0.00	0.85	0.96	0.98	0.99
Q	0.99	0.09	0.02	0.01	0.00

or conductance. It should be noted however, that $\bar{K}_{inter}$ is less responsive to increases in inter-cluster connectivity than $\bar{K}_{intra}$ is to increases in intra-cluster connectivity and that a graph's overall connectivity ($\bar{K}$) closely reflects inter-cluster connectivity, especially in cases where the number of clusters is large. Additionally, we note modularity and conductance display very counterintuitive behaviors, although on a much larger scale. In the following sections we attempt to explain these unintuitive behaviors and explain why the "Kappas" provide a more accurate picture of the graph's and clusters' connectivity patterns than either modularity or conductance.

Finally, in the following sections, we also show that the erratic behavior displayed by modularity and conductance are the result of their sensitivity to numerical implementation and sample sizes. This numerical sensitivity deeply affected our results with our moderately-sized graphs and clusters. As we will show in the next sections, this numerical sensitivity would only be compounded in the case of a larger data set, rendering these measures even less responsive. These sensitivities to data set size are particularly relevant in the context of large data sets ("big data").

4.1 Modularity Under Stress Test

In order to illustrate the lack of responsiveness of modularity and explain the results in the previous section, we examine the following numerical example: $|C| = 200, N = 16,400$ and $n_i = 82\ \forall i$. We then adjust the intra and inter-cluster connectivities, to examine the effect on modularity. The results are shown in Tables 5 and 6.

We begin with a clustering algorithm that would be very deficient and returns "clusters" that have 0% connection within themselves but are fully connected to the rest of the graph ($A0$). We gradually increase intra-cluster connectivity to 25% ($A25$) and 100% ($A100$), while keeping inter-cluster connectivity constant at 100%. We then do the opposite, we begin with 200 isolated complete graphs (in $B0$, each cluster is an isolated complete graph) and then increase inter-cluster connectivity to 25% ($B25$). These experiments are almost the same as those shown in Sect. 3, except that we kept cluster size constant, at 82 vertices, in order to facilitate calculations.

Table 5. Varying intra-cluster connectivity

Scenarios	A0		A25		A100	
Components of Q	e_ii	a_i	e_ii	a_i	e_ii	a_i
cluster 1	0	0.005	0.00001	0.005	0.00002	0.005
cluster 2	0	0.005	0.00001	0.005	0.00002	0.005
$\vdots$	$\vdots$	$\vdots$	$\vdots$	$\vdots$	$\vdots$	$\vdots$
cluster K	0	0.005	0.00001	0.005	0.00002	0.005

In Table 5, we see that with NO connectivity within clusters, $Q \approx 200 \times (0 - 0.005^2) \approx 0$. Now if we raise the intra cluster connectivity from 0% to 25%, we add $\lceil 0.25 \times 82 \times 81 \rceil = 831$ edges to the graph, all of which connect vertices within clusters.

The a_i portion remains essentially unaffected, because the a_i of each node is scaled by $\frac{1}{4m^2}$ (i.e., increase of $\frac{831}{4m^2}$). On the other hand, $e_{i,i}$, which is scaled by $\frac{1}{2m}$ (i.e., increase of $\frac{831}{2m}$) goes up ever so slightly, but on a different order of magnitude, and the denominator (m) also increases. So in the end, the added connectivity only has an infinitesimal effect on the value of Q:

$$Q \approx 200 \times (0.00001 - 0.005^2) \approx 0$$

Increasing the intra-cluster connectivity even further to 100% does not affect the value of Q either. Indeed, the number of intra-cluster edges increases to $82 \times 81 \times 0.5 = 3,321$, but this increase is scaled by $\frac{1}{2m}$ or $\frac{1}{4m^2}$, while m also increases as well. So in the end, Q remains indistinguishable from 0, $Q \approx 200 \times (0.00002 - 0.005^2) \approx 0$.

Table 6. Varying inter-cluster connectivity

Scenarios	BO		B25	
Components of Q	e_ii	a_i	e_ii	a_i
cluster 1	0.005	0.005	0.00038	0.005
cluster 2	0.005	0.005	0.00038	0.005
⋮	⋮	⋮	⋮	⋮
cluster K	0.005	0.005	0.00038	0.005

In Table 6, we observe that when none of the vertices within clusters are connected to vertices outside their cluster, yet all have connections to vertices within their assigned clusters (case of K isolated complete graphs), $e_{i,i} = a_i$. As a result $Q \approx 200 \times (0.005 - 0.005^2) \approx 1$. But as soon as inter-cluster connectivity increases, Q collapses. Increasing inter-cluster connectivity dramatically increases m, which dramatically reduces $e_{i,i}$. Simultaneously, a_i increases, although very modestly. With 200 connected components, modularity quickly reaches its maximum, $Q \approx 200 \times (0.005 - 0.005^2) \approx 1$. With 25% inter-cluster connectivity, it quickly approaches 0, $Q \approx 200 \times (0.00038 - 0.005^2) \approx 0.07$. Note that although the degree of each vertex does indeed increase and contribute to increasing each a_i, the denominator of each a_i is $4m^2$, a graph-wide number. In the end, any increase in the cluster-centric numerator of a_i is eliminated by a dramatic graph-wide increase in m. Also note that, predictably, increases in inter-cluster connectivity beyond 25% make Q rapidly converge to zero.

4.2 Conductance Under Stress Test

Conductance is calculated at the cluster level and we assign $\Phi(G)$ the minimum value of all $\phi(S)$. Taking the minimum makes conductance very sensitive to outliers and not robust at all. In the event the graph has even one single cluster, call it $\tilde{S}$, that is densely connected, then $\phi(\tilde{S}) \approx 0$. Consequently, $\Phi(G) \approx 0$, regardless of network configuration.

In the results shown in Sect. 3, conductance breaks down for a different reason, however: In the case of an edge-less graph the denominator of conductance is zero, so we set $\phi(S) = 0$, by convention. Later, as we raise intra-cluster connectivity, the denominator remains zero, because inter-cluster connectivity is kept at 0% (Table 1). In the case of completely disconnected "clusters" (incorrectly labeled as clusters by the algorithm), the denominator is again 0. The denominator remains unchanged, when we increase inter-cluster connectivity (Table 2). This pattern repeats with the introduction of noise (Tables 3 and 4).

4.3   Kappas Under Stress Test

As shown in Sect. 3, our Kappas behave as expected, even if $\bar{K}_{inter}$ appears less responsive to graph structure than $\bar{K}_{intra}$, $\bar{K}_{inter}$ closely mirrors $\bar{K}$ and $\bar{K}_{intra}$ increases slowly in the case of our weighted examples. This relatively slow response and mirroring are completely consistent with the definitions. Note that when one edge is added anywhere on the graph, $\bar{K}$ goes up by $1/(0.5 \times N \times (N-1))$, a very small amount. When one edge is added within a cluster, $\bar{K}_{intra}$ also goes up, but by a larger amount:

$$(1/k)/(0.5 \times n_i \times (n_i - 1))$$

When an edge is added between clusters, $\bar{K}_{inter}$ also only goes up by a small amount:

$$\frac{\frac{1}{0.5 \times \kappa \times (\kappa - 1)}}{0.5 \times [(n_i + n_j)(n_i + n_j - 1) - n_i(n_1 - 1) - n_j(n_j - 1)]}$$

In the case of weighted graphs, our weights $(w_{i,j})$ are all in the $[0, 1]$ interval, so when one edge is added within a cluster $\bar{K}_{intra}$ increases by

$$(w_{i,j}/k)/(0.5 \times n_i \times (n_i - 1)) \le (1/k)/(0.5 \times n_i \times (n_i - 1)).$$

It is also important to note that even in instances where $\bar{K}_{inter}$ or $\bar{K}_{intra}$ are not as responsive as expected, the relative magnitude of the measures still correctly identifies highly clustered graphs. In all our experiments strong clusters were always characterized by the inequality $\bar{K}_{intra} > \bar{K} > \bar{K}_{inter}$.

Finally, we call the readers' attention to the standard errors of the various Kappas, which remain stable around 0. We show standard errors to emphasize the statistical nature of the Kappas. However, due to the pre-defined homogeneous connectivity patterns used in our computational experiments, variance (standard deviation) in connectivity is relatively low. Additionally, a small standard deviation is then scaled by a relatively large denominator ($\sqrt{200}$), which reduces it even more.

4.4   An Example of Formal Statistical Testing for Kappas

As discussed previously, one of the strengths of our measures is their statistical definition. This definition allows us to perform formal statistical testing to confirm our conclusions. Here, we illustrate our claim by showing two examples, in Table 7. Our null hypotheses are, in the first test, $\bar{K}_{intra} \le \bar{K}$ and, in the second test, $\bar{K}_{inter} \ge \bar{K}$. The goal of these tests is to formally verify the quality of the clustering identified by an algorithm. If the clustering is good, the null hypotheses $\bar{K}_{intra} \le \bar{K}$ and $\bar{K}_{intra} \ge \bar{K}$ should be rejected, at the usual confidence levels $(0.01, 0.05)$. If the clustering is bad, as it is in our first example, we expect the null not to be rejected.

Table 7. Hypothesis test example

	Test $\bar{K}_{intra}$	Test $\bar{K}_{inter}$
Null Hyp	$\bar{K}_{intra} \leq \bar{K}$	$\bar{K}_{intra} \geq \bar{K}$
Alt. Hyp	$\bar{K}_{intra} > \bar{K}$	$\bar{K}_{inter} < \bar{K}$
Pct inter (actual)	1	0.75
Pct intra (actual)	0.75	1
$|C|$	200	200
$\bar{K}$	1.00	0.56
$\bar{K}_{intra}$	0.74	na
Std Error	0.01	na
$\bar{K}_{inter}$	na	0.54
Std Error	na	0.001
t-statistic	-26	-20
Deg freedom	199	19,899
p-value	0.000	0.000
Reject null?	**NO**	**YES**

5   Conclusion

We described a new set of statistical clustering measures that allow formal quality assessments and comparison of algorithms. Our measures are shown to be more robust than the commonly used modularity and conductance. In particular, our measures appear to be more responsive to cluster labeling and less sensitive to sample size, resolution limit and breakdowns during numerical implementation. This latter feature is especially important in the context of larger data sets.

In this article, we restricted our attention to non-overlapping clusters, since that is what most clustering techniques identify. Future investigations could focus on extensions to measuring the strength of overlapping clusters.

Acknowledgements. PM thanks Liudmila Ostroumova Prokhorenkova, Mark Newman, Cris Moore, Aaron Clauset and Anne Morvan, for their helpful comments and guidance. PM was supported by Mitacs-Accelerate PhD award IT05806.

References

1. Almeida, H.M., Guedes, D.O., Meira Jr., W., Zaki, M.J.: Is there a best quality metric for graph clusters? In: Machine Learning and Knowledge Discovery in Databases - European Conference, ECML PKDD 2011, Athens, Greece, 5–9 September 2011, Proceedings, Part I, pp. 44–59 (2011)

2. Aloise, D., Caporossi, G., Hansen, P., Liberti, L., Perron, S., Ruiz, M.: Modularity maximization in networks by variable neighborhood search. In: Bader, D.A., Meyerhenke, H., Sanders, P., Wagner, D. (eds.) Graph Partitioning and Graph Clustering, 10th DIMACS Implementation Challenge Workshop, Georgia Institute of Technology, Atlanta, GA, USA, 13–14 February 2012, Proceedings, pp. 113–128 (2012). http://www.ams.org/books/conm/588/11705

3. Biswas, A., Biswas, B.: Defining quality metrics for graph clustering evaluation. Expert Syst. Appl. **71**, 1–17 (2017). http://www.sciencedirect.com/science/article/pii/S0957417416306339

4. Brandes, U., Delling, D., Gaertler, M., Gorke, R., Hoefer, M., Nikoloski, Z., Wagner, D.: On modularity clustering. IEEE Trans. Knowl. Data Eng. **20**(2), 172–188 (2008). https://doi.org/10.1109/TKDE.2007.190689

5. Clauset, A., Newman, M.E.J., Moore, C.: Finding community structure in very large networks. Preprint **70**(6), 066111 (2004)

6. Creusefond, J., Largillier, T., Peyronnet, S.: On the evaluation potential of quality functions in community detection for different contexts. ArXiv e-prints, October 2015

7. Djidjev, H., Onus, M.: Using graph partitioning for efficient network modularity optimization. In: Bader, D.A., Meyerhenke, H., Sanders, P., Wagner, D. (eds.) Graph Partitioning and Graph Clustering, 10th DIMACS Implementation Challenge Workshop, Georgia Institute of Technology, Atlanta, GA, USA, 13–14 February 2012, Proceedings, pp. 103–112 (2012). http://www.ams.org/books/conm/588/11713

8. Fortunato, S.: Community detection in graphs. Phys. Rep. **486**, 75–174 (2010)

9. Fortunato, S., Barthélemy, M.: Resolution limit in community detection. Proc. Nat. Acad. Sci. **104**(1), 36–41 (2007). http://www.pnas.org/content/104/1/36.abstract

10. Holder, L.B., Caceres, R., Gleich, D.F., Riedy, J., Khan, M., Chawla, N.V., Kumar, R., Wu, Y., Klymko, C., Eliassi-Rad, T., Prakash, A.: Current and future challenges in mining large networks: report on the second SDM workshop on mining networks and graphs. SIGKDD Explor. Newsl. **18**(1), 39–45 (2016). http://doi.acm.org/10.1145/2980765.2980770

11. Huang, H., Liu, Y., Hayes, D., Nobel, A., Marron, J., Hennig, C.: (15) Significance testing in clustering. In: Hennig, C., Meila, M., Murtagh, F., Rocci, R. (eds.) Handbook of Cluster Analysis, pp. 315–335. Chapman and Hall/CRC (2015)

12. Lancichinetti, A., Fortunato, S., Radicchi, F.: Benchmark graphs for testing community detection algorithms. Phys. Rev. E Stat. Nonlinear Soft Matter Phys. **78**, 046110 (2008)

13. Leskovec, J., Lang, K.J., Mahoney, M.W.: Empirical comparison of algorithms for network community detection. ArXiv e-prints, April 2010

14. Leskovec, J., Lang, K.J., Dasgupta, A., Mahoney, M.W.: Statistical properties of community structure in large social and information networks. In: 7th International Conference on WWW (2008)

15. Morvan, A., Choromanski, K., Gouy-Pailler, C., Atif, J.: Graph sketching-based massive data clustering. In: SIAM International Conference on Data Mining (SDM 2018) (2018, to appear)

16. Moschopoulos, C.N., Pavlopoulos, G.A., Iacucci, E., Aerts, J., Likothanassis, S., Schneider, R., Kossida, S.: Which clustering algorithm is better for predicting protein complexes? BMC Res. Notes **4**(1), 549 (2011), https://doi.org/10.1186/1756-0500-4-549

17. Newman, M.E.J., Girvan, M.: Finding and evaluating community structure in networks. Phys. Rev. E Stat. Nonlinear Soft Matter Phys. **69**, 026113 (2004)

18. Ostroumova Prokhorenkova, L., Prałat, P., Raigorodskii, A.: Modularity of complex networks models. In: Bonato, A., Graham, F.C., Prałat, P. (eds.) Algorithms and Models for the Web Graph, pp. 115–126. Springer, Cham (2016)
19. Ostroumova Prokhorenkova, L., Prałat, P., Raigorodskii, A.: Modularity in several random graph models. Electron. Notes Discrete Math. **61**, 947–953 (2017). http://www.sciencedirect.com/science/article/pii/S1571065317302238.   The European Conference on Combinatorics, Graph Theory and Applications (EUROCOMB 2017)
20. Reichardt, J., Bornholdt, S.: When are networks truly modular? Physica D Nonlinear Phenom. **224**(1), 20–26 (2006). http://www.sciencedirect.com/science/article/pii/S0167278906003678. Dynamics on Complex Networks and Applications
21. Sanders, P., Schulz, C.: High quality graph partitioning. In: Bader, D.A., Meyerhenke, H., Sanders, P., Wagner, D. (eds.) Graph Partitioning and Graph Clustering, 10th DIMACS Implementation Challenge Workshop, Georgia Institute of Technology, Atlanta, GA, USA, 13–14 February 2012, Proceedings, pp. 1–18 (2012). http://www.ams.org/books/conm/588/11700
22. Spielman, D.A., Teng, S.H.: A local clustering algorithm for massive graphs and its application to nearly linear time graph partitioning. SIAM J. Comput. **42**(1), 1–26 (2013)
23. Yang, J., Leskovec, J.: Defining and evaluating network communities based on ground-truth. CoRR abs/1205.6233 (2012). http://arxiv.org/abs/1205.6233
24. Yang, J., Leskovec, J.: Overlapping community detection at scale: a nonnegative matrix factorization approach. In: WSDM 2013. ACM, 978-1-4503-1869-3/13/02 (2013)

Author Index

Printed in the United States
By Bookmasters